# INFOSENSE

## Turning Information into Knowledge

Keith Devlin

W. H. Freeman and Company
New York

*Text Design:* Cambraia Fernandes

Cataloging-in-Publication Data available from the Library of Congress.

ISBN 0-7167-3484-2 (hardcover) ISBN 0-7167-4164-4 (paperback)

Printed in the United States of America

First paperback printing, 2001

W. H. Freeman and Company
41 Madison Avenue, New York, NY 10010

# Contents

# PREFACE ------------------------------------------◀

In some ways, you can think of this book as a "how to survive" manual for life in the knowledge age. It is based in large part on fifteen years of research into the nature of information carried out at Stanford University's Center for the Study of Language and Information (CSLI), an academic think tank established in 1983 with funds from the System Development Foundation (a spin-off from the RAND Corporation). Since 1985, I have played an active role in that research project.

Until now, the many fundamental scientific results emanating from CSLI have been confined largely to the academic research community, available only in technical research papers written for other experts. This book brings to the general reader and to the business community some of the key results that have been obtained at CSLI, and it shows how those results can be used to improve the way companies (and individuals) manage information.

In 1991, I published the results of my research, and that of my collaborators at CSLI, in the book *Logic and Information* (Cambridge University Press). That book was awarded the American Association of Publishers' prize of Most Outstanding Book in Computer Science and Data Processing of 1991.

In *Logic and Information* I explained the basic ideas of a new theory of information in a straightforward and relatively accessible manner. As a result, despite its technical content matter, the book won acclaim from some members of the business community. Since its appearance, I have occasionally been asked to consult on matters relating to information management. Some of the businesspeople whom I have encountered suggested that I should write a version of the book that directly addressed their needs. This is that book. In the pages that follow, I draw upon my consulting experiences to show how to take ideas of the new theory—the new science, if you will—and put them to real use.

Some of the material I present in this book has also been discussed, in different forms and for different audiences, in two other books I have written, in addition to *Logic and Information*. Those books are *Language at Work* (with Duska Rosenberg, 1996) and *Goodbye, Descartes* (1997). Since all four books—the present volume plus the three just mentioned—are drawing on the same central body of research, it is inevitable that, in places, the exposition in two or more of the books is very similar. Moreover, over the years, I have developed what I think are "the best ways" to get particular points across. So readers of this book who are familiar with my earlier books on the subject of information will occasionally experience a feeling of déjà vu.

Despite the similarities in places, however, all four books are different in significant ways—in content, in aim, and in the intended audience. In many ways they complement one another. My aim is to convey certain information to you, the reader, in the most efficient way I can find. For the purpose of scholastic honesty, however, let me list here the places where there is a significant overlap with one of my other books.

Parts of the Prologue and of Chapter 2 are adapted from *Language at Work*. Some passages in Chapter 3 are adapted from *Goodbye, Descartes*. In Chapters 4 and 6 I use some of my favorite examples, all of which I included in *Logic and Information* and in *Goodbye, Descartes*. Some sections of Chapter 8 are adapted from *Goodbye, Descartes*. Chapters 9 and 20 present, respectively, an account of the work of Paul Grice on conversation and of Hubert and Stuart Dreyfus on expertise. Similar accounts appear in *Goodbye, Descartes*.

On the face of it, this book is written for the professional businessperson. Any author has to have an ideal reader in mind, and in order to write this book I found it convenient to think of a person working in business: the CEO, the middle manager, the ambitious young assistant, the office worker, the sales and marketing personnel, or the telecommuting professional. But the nature of the knowledge age is that the same message is just as relevant to anyone. Accordingly, though most of my examples are drawn from the business world, and though most of the suggestions I make are ostensibly targeted at the businessperson, I have tried to keep my account at a level that is accessible to anyone.

In addition to presenting a close look at some major applications, I include numerous, everyday, real-life examples, and give specific, step-by-step advice on how to improve the efficiency of information management, based on sound scientific principles. Simple, clear diagrams help explain the basic ideas and provide useful business management tools. Each chapter is short, enabling the reader to tackle it in short bursts, and ends with a brief summary.

## ABOUT THE AUTHOR

For those readers (I am one of them) who like to know a little of the backgrounds of their authors, here are a few details: I trained as a mathematician, obtaining my Ph.D. in mathematics at the University of Bristol in 1971. After a successful career in mathematical research, in the early 1980s I switched my attention to the investigation of information that led to my involvement with CSLI. I do, however, continue to be involved in mathematics and mathematics education, and, since ceasing to carry out fundamental research in mainstream mathematics (in order to try to understand information), I have written a number of popular books on mathematics, most recently *Life by the Numbers* (John Wiley, 1998), written to accompany the PBS television series of the same name, and *The Language of Mathematics: Making the Invisible Visible* (W. H. Freeman, 1998).

Among a current total of twenty-two books and one interactive CD ROM, the others I have written for the general reader are *Goodbye, Descartes: The End of Logic and the Search for a New Cosmology of the Mind* (John Wiley, 1997), *Mathematics: The New Golden Age* (Penguin, 1987; Columbia University Press, 1999), and *Mathematics: The Science of Patterns* (W. H. Freeman, Scientific American Library, 1993, 1996).

The new "theory of information" I describe in this book began with the work of two Stanford University researchers at the end of the 1970s: Jon Barwise and John Perry. Respectively a mathematician and a philosopher, Barwise and Perry set out to develop a new theory of information based on sound mathematical principles. Both of these two fine researchers have offered me considerable support in my own investigation of information, and I owe much of my present understanding of information to many conversations with them.

The research project Barwise and Perry began received a considerable boost in 1983, when the System Development Foundation donated $23 million to a consortium consisting of Stanford University, Stanford Research Institute (SRI), and Xerox PARC to establish at Stanford the Center for the Study of Language and Information (CSLI), a multidisciplinary research institute with the goal of developing a basic science of information. Barwise was the first director, Perry the second some years later. I spent the two years 1987 to 1989 as a full-time researcher at CSLI and have retained my association with them ever since. I am currently a Senior Researcher at the center.

In addition to my position as a Senior Researcher at CSLI, I am Dean of Science at Saint Mary's College of California in Moraga, California, and a Consulting Research Professor in the Department of Information Science at the University of Pittsburgh.

Conversations and collaborations with various colleagues have contributed to the ideas that have found their way onto the pages of this book. In particular, I should mention Jon Barwise (Indiana University), Lee Bloomquist (Steelcase, Inc.), John Etchemendy (Stanford University), Ted Goranson (Sirius-Beta), David Israel (SRI International), David Leevers (independent consultant), John Perry (Stanford University), Stanley Peters (Stanford University), and Duska Rosenberg (Brunel University).

John Michel, my editor at W. H. Freeman, who was enthusiastic about the project from the start, provided me with numerous suggestions as to how to make this book more widely accessible.

I can be reached by mail at School of Science, Saint Mary's College of California, Moraga, CA 94575, by fax at (925) 631-7961, and by e-mail at **devlin@stmarys-ca.edu** or **devlin@csli.stanford.edu**.

In early human history, technological advantages were built on the availability of certain plants, animals and geographies.

In today's emerging information society, the critical natural resources are human intelligence, skill and leadership. Every region of the world has these in abundance, which promises to make the next chapter of human history particularly interesting.

*William E. Gates*
*CEO, Microsoft Corporation*
*The Reality Club, April 1998*
*http://www.edge.org*

# From Innocence to InfoSense

---

In recent years there have been any number of attempts to improve information flow in business. For a period, the buzzword was "information." Then, starting with the appearance of the 1995 book *The Knowledge-Creating Company*, by two Japanese academics Ikujiro Nonaka and Hirotaka Takeuchi, the word "information" has tended to be replaced by "knowledge." To a large extent, the issues have remained the same, however. The only change of any significance has been an increased awareness of the crucial role played by individual human psychology and by culture. As Nonaka and Takeuchi observe, "knowledge, unlike information, is about beliefs and commitment."

In this book, however, I regard the distinction between information and knowledge as a fundamental one, to be investigated in a theoretical fashion. In this regard, the present work is fairly unusual among attempts to investigate information or knowledge.

## THE PLOUGHSHARE OF TOMORROW

The past twenty years have seen a major transformation in the nature of human life, at least in a large part of the world—what we generally refer to as the "civilized world." We have moved from an industry- and transportation-based society to a society based on knowledge and information. (The exact relationship between knowledge and information is one of the things we'll explore in this book.

Though closely related, these concepts are not the same. Roughly speaking, knowledge is information that a person possesses in a form in which he or she can make immediate use.)

These days, knowledge and information are what bind us together, and many people now earn their entire living acquiring and transferring knowledge or processing and handling information in one way or another.

That heavy dependency on knowledge and information is going to increase in the years to come.

Already today and especially tomorrow, a basic understanding of information will be as important as were farming skills in the agricultural age or basic industrial skills in the industrial age.

In fact, an understanding of information—what it is, how it flows, how to acquire and use it, and what it takes to turn information into knowledge—will soon be more important to the average citizen than was a knowledge of industrial techniques in the industrial age. Let me elaborate.

In the middle ages, many people needed to know how to grow crops and tend animals, since that was how they made their living. But when the industrial age came along, production became concentrated in the hands of just one sector of the community. Only that portion of the population needed the basic industrial skills; others in society performed other functions, requiring different knowledge. As we enter the knowledge society, we are entering an era much more like the agricultural age. Though development and manufacture of the key information processing technologies are concentrated in just a few hands, the nature of that technology is to put the tools of information handling into everybody's hands. In the twenty-first century no citizen will be able to function properly without a basic understanding of information and an appreciation of what is required to turn information into knowledge. Knowledge will be the ploughshare of tomorrow.

## WHAT IS INFORMATION?

But what exactly do we mean by knowledge? What is information? How is information stored? And what is required to turn information into knowledge? Everyone thinks they know the answers to these questions until they are asked to say what those answers are. At that point, it becomes clear that we have at best a vague idea of what the words mean. Moreover, your vague idea and mine may not be the same. This is quite unlike the situation in, say, civil, mechanical, or electrical engineering.

The engineers who produce our buildings, bridges, automobiles, aircraft, household appliances, and communications technologies base their work on the solid foundation of decades and even hundreds of years of scientific progress in physics and other sciences. But the people who design and manage our information systems (which may comprise people or machines or a combination of the two) have to work with much shakier foundations.

The fact is, when you stop to compare the scientific foundations of, say, civil, mechanical, or electrical engineering with the theoretical understandings on which we base the ways we handle information, you realize that, when it comes to information and knowledge, we are still innocents. And yet, modern life depends heavily on information and knowledge. If we handle it poorly, because we do not really understand it, then we will always be just a step away from disaster. We need to learn how to handle this stuff called "information" in a safe and efficient manner. Our present information innocence reminds me of Marie Curie, who handled radioactive uranium in her laboratory on a daily basis, unaware of the dangers it posed for her. Marie Curie paid for her scientific innocence with her life. In our daily handling of information, we need to progress from innocence to *info-sense*. And we need to do so before it is too late.

In this book I summarize one attempt to develop a scientific understanding of information and knowledge. I shall begin by explaining the results of twenty years of fundamental scientific research into the nature of information: what it is, how it arises, how it can be stored, and how it can be transmitted. By regarding information as a substance, we can study it mathematically, using methods very much like those used in the natural sciences.

I will go on to show how that scientific understanding of information can be used to improve the way businesses—and individuals—manage information.

Once we have a scientifically based understanding of information, we can then turn to the question of what it takes to turn information into knowledge. For in the end, it is not information that we use, but the knowledge that we get as a result of obtaining that information.

## KNOWLEDGE—BY THE PEOPLE, FOR THE PEOPLE

Though we often speak of information being "a valuable commodity," the value of information lies in its potential to be turned into knowledge. For, ultimately, knowledge is what makes the difference in what we can do, and the value of information depends upon the value of the knowledge to which it can lead.

The early chapters of this book are devoted to a scientific analysis of information, but knowledge is our real goal. I begin with a study of information because the better our understanding of information, the greater is our ability to turn information into knowledge and to use knowledge more effectively.

Since information may be regarded as a *substance* having a certain *structure*, we may study it in a mathematical fashion, independent of its possession or use. (This is why you are reading a book on information and knowledge written by a mathematician!) In contrast, knowledge is information put into practice—or at least possessed in a form that makes it immediately available to be put into practice. In particular, knowledge requires a knower. Consequently, a study of knowledge is properly carried out using the methods of the human sciences, in particular psychology, sociology, and management science. Thus, when we come to study how to turn information into knowledge, we shall find ourselves adopting a different approach. In particular, because knowledge is largely a matter of human practice, the discussion of knowledge tends to be much more anecdotal than the treatment of information that precedes it.

It is mostly in its use of a scientific theory of *information* that this book makes a novel and hitherto unique contribution to the literature. In contrast, there are a number of excellent books that discuss knowledge. One that I draw on in particular is Thomas Davenport and Laurence Prusak's excellent *Working Knowledge,* published in 1998.

## INFOSENSE

Although I talk about a *science* or *theory* of information, for the most part I shall leave the fine details out of this account. My intention is to provide you with a highly practical "how to" guide to information and knowledge. In our book *Language at Work,* Duska Rosenberg and I show how to take some of the scientific ideas outlined in this book and apply them in a highly technical fashion; here,

I describe the science in order to help you develop a *commonsense understanding* of information—what I am calling "infosense." Only the information systems expert needs the level of scientific understanding required to carry out a detailed analysis of, say, a document, as described in *Language at Work;* but every one of us needs a reasonable level of infosense to survive and prosper in today's world.

Just because my account is written at the level of everyday common sense, however, it does not follow that the underlying scientific theory is unimportant. On the contrary, it is only by starting with a sound scientific understanding of information that we can hope to develop a reliable information sense.

Of course, for most purposes, all we need is an intuitive acquaintance with the underlying theory. The same is true in other areas. For example, the baseball pitcher's skill is based on sound principles of physics, but how many players understand gravitational theory and the aerodynamics of projectiles? Likewise, using electrical appliances depends on the physics of electric current, but for most of us it's enough to know about the danger of live wires and the importance of grounding circuits. When it comes to information, what most of us need most of the time is infosense, not a theory of information. And what infosense amounts to is a somewhat superficial, intuitive acquaintance with the theory of information.

It is because it is built on a sound, theoretical investigation of information that this book differs from the majority of business books with the word "information" or "knowledge" in their titles that you will find on the shelves of your local bookstore.

Many of those books have no theoretical basis at all. (A friend of mine, a senior figure in the United Kingdom business world, observing that such books are always sold in large numbers at airport bookstores, refers to them derisively as coming from "the Heathrow School of Management.")

Some books, particularly those written by successful CEOs, are based on the author's hunches. That approach can sometimes work, occasionally spectacularly, particularly if the person with the hunch has remarkable insight and intuition. I suspect, however, that luck generally plays a major role. For real progress, however, we need more than hunches. We need a solid science on which to build our infosense: a science of information.

In this book, I outline the basic science and give some applications—real applications to real problems, both in everyday life and in real companies.

# SITUATION THEORY

The theoretical framework I shall use to investigate information is called "situation theory." It was developed by Barwise and Perry in the early 1980s and described in my book *Logic and Information.**

Situation theory starts out by recognizing that information is transmitted in situations—for example, when one colleague talks to another, when a person sits down at a computer and consults a database, or when one computer communicates with another. In order to analyze the way information flows, you have to begin by examining those kinds of situations. (Exactly why you have to look at the situation will be made clear in the pages that follow.) Situation theory sets out to analyze situations just as atomic physics sets out to analyze atoms.

In the coming chapters, I will outline situation theory in simple, straightforward terms. In the meantime, let me remark that, although the theory is still very much in its infancy, it has already chalked up a number of specific—and, in some cases, spectacular—successes, among them:

◆ The resolution of the liar paradox of Epimenedes, a philosophical puzzle about language that was first posed by the ancient Greeks, and which had resisted numerous attempts at solution. The liar paradox arises when a man stands up and says, "What I am now saying is false." If you examine this statement, you find it is paradoxical. If the man is telling the truth, then he is lying; if he is lying, then he is telling the truth. The liar paradox highlights problems that can arise when you have self-reference. This has practical implications, since most computer programs can refer to themselves, for instance by calling themselves as subroutines. [For the solution, see Barwise & Etchemendy (1987).]

◆ The design and implementation of *Hyperproof,* an interactive instructional computer program to develop skills in formal reasoning using information presented in a combination of linguistic and visual forms. [See Barwise & Etchemendy (1994).]

◆ The first realistic and useful analysis of "common knowledge," a fascinating phenomenon that plays a significant role in social, business,

---

*An earlier account can be found in Barwise and Perry's own book *Situations and Attitudes,* but the theory underwent several major changes in the years immediately following the appearance of that first book on the subject, and in many ways *Logic and Information* is an update of the Barwise and Perry book.

political, and military life. Common knowledge occurs when two or more people not only know something but also all know that they all know, they all know that they all know that they all know, and so on. [See Barwise (1989), Chapter 9.]

◆ An in-depth analysis of the common linguistic device of "anaphora," whereby pronouns and other parts of speech make reference to things in the world by utilizing the prior appearance of certain words or phrases within the phrase, sentence, or discourse. For example, if I say "John fell down. Mary pushed him." my use of "him" in the second sentence clearly refers to John. We know this because the word "John" has already appeared in the first sentence. [See Gawron & Peters (1990).]

◆ A systematic treatment of counterfactual statements providing a resolution of some long-standing problems with understanding how such statements work linguistically. Counterfactual statements are of the form "If A, then B," where A is either false or else its truth or falsity is not known, possibly because it refers to some future event. [See Barwise (1989), Chapter 5.]

◆ A comprehensive analysis of "indexicality," the feature of language whereby the same words or phrase can mean different things, depending on the context, in particular on who is speaking, where, and when. [See Barwise & Perry (1983), Devlin (1991).]

◆ A method for modeling business and manufacturing processes that has already found commercial use. [See Menzel & Mayer (1996), Devlin (1996).]

◆ A mathematically based analysis of the manner in which cultural knowledge influences the way we use and understand everyday language. [See Devlin & Rosenberg (1993).]

◆ An in-depth analysis of a particular instance of communication breakdown in the workplace, which led to a successful redesign of the data collection procedures used by the company concerned. [See Devlin & Rosenberg (1996). See also Chapter 5 of this book.]

◆ A study of the effect of different office layouts on the efficiency and reliability of information flow among knowledge workers. [See Devlin (1997b).]

All of those early successes, obtained within twenty years of the birth of the new theory, were a result of the degree to which the theory came to grips with the truly fundamental questions about information and communication. That is where we need to start our journey toward successful information management.

# 1

# The Grin of the Cheshire Cat

----------------------------------------------------------------

I believe that a meaningful personal assessment of how infor-
mation-processing technology can and ought to shape our lives
must begin with an understanding of information itself and
what it means to "process" that substance. What is this raw
material that people and their machines use to make decisions?

*Arno Penzias,\* Head of Research, AT&T Laboratories,*
Ideas and Information, *Simon and Schuster, 1989, p. 37.*

## TWO LITTLE WORDS

A pleasant climate makes the Canary Islands a popular tourist destination all year round. Every weekend, planeloads of holiday-
makers fly in and out of the airports at Tenerife and Las Palmas.

On the morning of Sunday, March 27, 1977, however, the pleasant atmosphere was shattered when a terrorist bomb exploded in
the main shopping area of Las Palmas. As a precaution, the author-
ities diverted Las Palmas air traffic to Tenerife for the rest of the
day. That put a greater-than-usual strain on Tenerife's Los Rodeos
airport.

----

\*With his colleague Robert Wilson, Penzias is the discoverer of the oldest informa-
tion in the universe, the cosmic microwave radiation—the sole surviving trace of
the Big Bang with which the universe began. Penzias and Wilson were awarded the
1978 Nobel Prize in Physics for that discovery.

Adding to the problems, the weather that Sunday was poor, and by the afternoon the airport was covered in mist and rain. And, as if that weren't enough, the central lights on the airport's single main runway were not working, and only one of the airport's three main radio frequencies was operational, which meant that all flights had to share the same frequency.

At 5:00 p.m., after being delayed because of the weather and the overcrowded flight schedule, two Boeing 747 jumbo jets pushed back from the terminal and started their engines ready to taxi to the runway. One of the planes, belonging to the Dutch airline KLM, had arrived from Amsterdam earlier that day. The other, a Pan Am plane, had come in from Los Angeles. The captain of the KLM plane was Veldhuizen van Zanten; captain Victor Grubbs was in charge of the Pan Am plane. Both pilots were very experienced. The KLM plane was carrying 230 passengers and a crew of 14; the Pan Am plane was more fully loaded, with 364 passengers and a crew of 16.

The KLM plane was given taxiing instructions first. Van Zanten was told to taxi across a link to the end of the runway, swing around, and prepare for takeoff. The Pan Am aircraft was to follow the KLM plane down the runway but turn off before the end onto a short diagonal link that would enable it to swing around behind the KLM plane and follow it into the air.

When the KLM aircraft was in position at the end of the runway, the pilot radioed to the tower: "We are ready for takeoff and are waiting for ATC [air traffic control] clearance."

The tower responded: "You are cleared to the Papa Beacon, climb to, and maintain, flight level nine zero [9,000 feet], right turn after takeoff, proceed on heading 040 until intercepting the 325 radial from Las Palmas VOR."

The Pan Am pilot then contacted the tower with his call sign: "Clipper 1736."

*Tower:* Papa Alpha 1736, report runway cleared.
*Pan Am:* We'll report runway cleared.
*Tower:* Okay, thank you.

Seconds later, as the Pan Am plane was still taxing down the main runway, the KLM flight rammed into it at takeoff speed, killing everyone on board the KLM plane and all but 54 on board the Pan Am jumbo. A total of 583 deaths, the worst airline disaster ever.

The first officer on the Pan Am flight, Robert Bragg, was among those who survived. He described the final moments before the crash. "We saw lights ahead of us in the fog. At first we thought it was the KLM aircraft standing at the end of the runway. Then we realized they were coming towards us."

Bragg shouted into the radio: "Get off! Get off!" Captain Grubbs called to the tower: "We're still on the runway" and made a desperate attempt to swing his plane off the runway onto the grass.

By then, the KLM aircraft had its nose in the air, and so the crew would have been unable to see the Pan Am plane in its path. A few seconds later the KLM plane became airborne, before colliding with the roof of the Pan Am aircraft, tearing it open. It then crashed back onto the tarmac, slid 600 yards along the runway, and exploded. There was a slight delay before the Pan Am plane too caught fire, just enough time for the 54 survivors to escape.

None of this was visible from the tower. The controllers were unaware that anything had happened until another aircraft reported seeing a fire on the runway. By the time the emergency services were dispatched, it was too late to save any more lives.

Though the overcrowding at the airport, the inadequate runway facilities, and the weather all set the scene for the disaster, it seems clear that the crucial error was one of miscommunication. The most likely sequence of events is this.

First, the KLM captain took the flight clearance instruction "You are cleared to the Papa Beacon" as a clearance for takeoff, which it was not. In addition, because both the KLM and Pan Am planes were sharing the same radio frequency, the KLM crew overheard the Pan Am pilot's message, "We'll report runway cleared." Unfortunately, they missed the two crucial words "we'll report," hearing instead only "runway cleared." As a result, they now thought both that they had been cleared for takeoff and that the Pan Am plane had cleared the runway. They never had an opportunity to correct this misunderstanding. Minutes later, 583 people were dead or dying.

Both aircraft worked perfectly. All of their sophisticated control and guidance systems were functioning normally. There were no mechanical failures, and although the weather played a role in the sequence of events that led up to the crash, the weather did not itself cause the crash. Rather, the critical failure was in the flow of information. Almost certainly, the loss of just one small piece of information—two words—was the principal cause of the Tenerife disaster. In all likelihood, had the KLM captain heard the Pan Am pilot say the two words "we'll report," both planes would have taken off safely as intended.

The fact is, in the complex society we live in today, small, seemingly simple pieces of information can sometimes assume enormous value. As we advance still further into the Knowledge Society and our dependency on information and on information technology becomes still greater, more and more aspects of our lives will depend on what seem to be mere shreds of information.

For our primitive ancestors, surrounded by dangerous predatory animals, life often depended on the ability to make and use tools and weapons. More recently, as modern humankind entered the industrial age, people's lives depended on the ability of a society to produce and transport fuel and food. But today, and much more so tomorrow, our lives and livelihoods will depend on our ability to process and transport information quickly and accurately.

To survive in the information society, anyone in a position of responsibility or authority—at any level—will need to understand what information is. They will need to know how it arises, how it flows, and how we can make sure we use it wisely.

The message you will find in this book is as relevant to the parent or teacher as it is to the airline pilot, to the college graduate struggling in a first job as to the seasoned CEO, and to the young rock musician as to the auto mechanic: No one who wants to survive for very long can afford to screw up the information handling they must perform every day.

So what exactly is information?

## WHAT IS INFORMATION?

You might think that we ought to be able to say what information is. After all, in an era of global communications, information is the thread that ties us all together. By being able to transmit vast amounts of information rapidly from continent to continent, we have transformed a widely separated and diverse world into a single global megalopolis. The messenger on foot or on horseback has given way to the Information Superhighway. So just what is this stuff called information?

Whatever it is, information can be a valuable commodity, to be collected, guarded, duplicated, sold, stolen, and sometimes killed for. Millions of people around the world spend their entire working day gathering, studying, and processing information. Entire industries have developed to manufacture equipment (and software) to store and process information.

You can't open a newspaper without reading the word "information." Numerous books have the word "information" in the title. Many people have the word "information" in their job titles.

With information all around us, then, you would think it was a simple task to say just what it is. But, like the Cheshire Cat in Lewis Carroll's *Alice in Wonderland*, when you start to look closely at this stuff called information, it seems to disappear before your eyes, leaving only a tantalizing grin.

Part of our difficulty in understanding information is that we are so used to dealing with information in our everyday lives—for instance, when we communicate with one another—that we often fail to see the complexities involved.

For example, when someone approaches us in the street and says "Do you have the time?", we look at our watch and tell them what the time is, without giving the exchange a moment's thought. And yet, taken literally, the question asked requires an answer of "yes" or "no." Why don't we take it that way? Why do we instead take the question, correctly, as a request to be told the time?

"Do you have the time?" is so familiar that it rarely causes problems in everyday communication. We all know the intended meaning of the question. At least, all native English speakers know this. Problems can, and do, arise when nonnative speakers first encounter such idioms. But even for native speakers, there can be problems with less ubiquitous phrases. The potential for misunderstanding is always present. Idiomatic and metaphorical use of language is not rare; it is the norm. Contrary to what we might expect, language is almost never used literally. In fact, some linguists argue that there is no such thing as literal meaning at all.

Given the importance of effective communication in business, if you are in business, whether you are a manager or a "drone," you need to find ways to reduce the possibility of error. To do this, you need to understand how communication takes place—how information is transmitted, from person to person, from machine to machine, and between person and machine. The majority of communication-related problems in business—some of them highly costly—can be traced back to an inadequate understanding of the nature of information.

The need to understand information—what it is and how it "flows"—is not restricted to the large company with a centralized "information processing unit." Whenever you have one person communicating with another, you have information flow, because communication can be regarded as a means for conveying information from one person to another. The communication can vary from making a simple statement, such as giving your name to someone you have just met, to providing someone with the plans for a new chemical-processing plant.

Other forms of communication can also be viewed as transmissions of information. For example, questions—quests for information—can be regarded as conveying information. If I ask you "What is the time?" I convey to you the information that I want to be told the time.

Persuasion also involves the transmission of information from one person or group of people to another. In fact, it has been estimated that fully one-quarter of the U.S. national product is linked solely to "persuasion."

## KNOWLEDGE ≠ INFORMATION

Though the words "information" and "knowledge" are often used interchangeably, the two are not the same. Nor is information the same as data, though those two terms are also often confused. Understanding the subtle distinctions between the three concepts of data, information, and knowledge is essential, and it will be a major focus in the pages ahead. But it will be useful to start out with some preliminary ideas.

Roughly speaking, data is what newspapers, reports, and "computer information systems" provide us with. For example, a list of stock prices on the financial page of a newspaper is data.

When people acquire data and fit it into an overall framework of previously acquired information, that data becomes information. Thus, when I read the list of stock prices in the newspaper, I obtain information about various companies. What allows me to acquire information from the data in the newspaper is my prior knowledge of what such figures mean and how the stock market operates.

The well-known management consultant and author Peter Drucker has described information as "data endowed with relevance and purpose." And academics and management consultants Thomas Davenport and Laurence Prusak say in their 1998 book *Working Knowledge* that "Data becomes information when its creator adds meaning" [p. 4].

In terms of an equation:

$$\text{Information} = \text{Data} + \text{Meaning}$$

When a person internalizes information to the degree that he or she can make use of it, we call it knowledge. For example, if I know how to buy and sell stocks and am familiar with some of the companies whose stock values are listed in the newspaper, the information I obtain by reading the figures can provide me with knowledge on which to trade stocks.

Davenport and Prusak define knowledge in this way:

> *Knowledge is a fluid mix of framed experiences, values, contextual information, and expert insight that provides a framework for evaluating and incorporating new experiences and information. It originates and is applied in the minds of knowers. In organizations, it often becomes embedded not only in documents or repositories but also in organizational routines, processes, practices, and norms.* [*Working Knowledge,* p. 5]

As an equation:

$$\text{Knowledge} = \text{Internalized information} + \text{Ability to utilize the information}$$

Information can be regarded as a "substance" that can be acquired, stored, possessed either by an individual or jointly by a group, and transmitted from person to person or from group to group. As a substance, information has a certain stability and is perhaps best thought of as existing at the level of society. Knowledge, on the other hand, exists in the individual minds of people. Since, as we all know, people appear much more complex and unpredictable at the individual level than does an entire society, it is not surprising that knowledge is somewhat harder to pin down than is information. As Davenport and Prusak go on to say immediately following the passage quoted above: "Knowledge exists within people, part and parcel of human complexity and unpredictability."

It seems, then, that information is not only more fundamental than knowledge, it might also be easier to get hold of. Accordingly, it is with information that we should begin our quest. With a scientifically grounded understanding of information under our belts, we can then turn to the question of knowledge: What exactly is required to turn information into knowledge, and how is it transmitted?

We need to be cautious, however. Using the precise tools of mathematics, we can understand and control with great accuracy the way information flows through computer networks. But how does information flow from person to person or between human and machine? The reason we have difficulty finding an answer to this question is this: When you turn from machines to people, the methods of mathematics no longer apply—at least not to anything like the same extent—and things look far less clear and precise.

With today's technology, we are so used to the success of the methods of science that we tend to think that science can bring progress wherever it is applied. When we apply science to the physical world, that is very often the case. Indeed, the methods of science were developed precisely with physical applications in mind. It is when we try to apply those same methods to human beings, and in particular to the human mind—how we think, how we communicate, and how we make decisions—that we run into difficulties.

To study the human mind and human behavior with anything like the degree of rigor and precision that is required to improve the way people work together, we almost certainly need a new scientific approach, that uses some new mathematical ideas.

## Summary

We are entering the knowledge society, a world where the principal fuel and the real currency that drives the economy is knowledge. Life in that society will require a basic understanding of what knowledge is, how it is created, and how it is transmitted from one individual to another.

A proper understanding of knowledge and information should be built on a solid scientific foundation.

One of the goals of the science of information presented in this book is to clarify the distinction between data, information, and knowledge. Roughly speaking, the distinction is this:

◆ Data is what you get when your computer prints out a table of figures or a list of names and addresses.

◆ Data becomes information when people acquire it in the course of their daily activities.

◆ Information becomes knowledge when a person internalizes it to a degree that it is available for immediate use.

◆ Data exists on paper and on computer disks.

- Information exists in the collective mind of a society.
- Knowledge exists in an individual person's mind.

If we want to understand knowledge and information, we should begin with information, which is both more fundamental and easier to pin down.

Since most of the analytic tools of science were developed to study the physical world, we should be prepared to develop new tools to study the more human domain of information flow.

# 2

# Getting Down to Business

------------------------------------------------------------

In a global economy, knowledge may be a company's greatest competitive advantage.

*Thomas Davenport & Laurence Prusak, 1998, p. 13.*

## THE PRODUCTIVITY PARADOX

In the Knowledge Society, information is everybody's business. But the business world is where you will find the cutting edge. A good indication that business managers have yet to come fully to grips with information technology can be found in what has become known as the productivity paradox: Since the 1950s, the United States has invested heavily in information technology, leading the rest of the world by a large margin. Yet this enormous investment has led to very little increase in productivity. In fact, the greater the investment in computers, the lower were the productivity gains. Consider the figures.

According to the U.S. census, in 1950 there were fewer than 900 computer operators in the United States, and in 1960 there were still only 2,000. By 1970 there were 125,000, and by 1985 around 500,000.

In that year, 1985, expenditure on information processing equipment accounted for 16 percent of total capital stock in the service sector (a total of $424 billion), up from 6 percent fifteen years earlier. In 1996, U.S. companies spent 43 percent of their capital budgets on

**19**

computer hardware—a colossal $213 billion, and more than they invested in factories, vehicles, or any other kind of durable equipment.

On top of the computer hardware costs were all the associated personnel costs, the software costs, the operating and maintenance costs, the costs of building and maintaining special air-conditioned computer rooms, and the associated training costs. Taken together, all of those additional costs amount to far more than the cost of the equipment. The total cost of computing for 1996 was about $500 billion in the United States and more than $1 trillion worldwide.

And, of course, while the costs were rocketing, so too was the computing power that each dollar bought, doubling roughly every 18 months.

How has this phenomenal growth in expenditure on computing affected productivity? Between 1950 and 1965, the early days of computing, U.S. productivity grew by about 2 percent per annum. Since the mid-sixties, as computer growth was accelerating, productivity gains have remained below 2 percent.

In short, the enormous growth in the use of computers has led to virtually no increase in productivity. This is the "productivity paradox."

That the growth in computing has transformed the workplace is beyond question. Many of us work in a very different fashion than we did before the growth in computing. Large numbers of people now spend their working days performing tasks that simply did not exist a half century ago. What these changes have not done, however, is lead to greater productivity. Increasing amounts of money are spent on "processing information," and even greater numbers of people spend their time "managing information," but when you check the bottom line you find there are no tangible gains.*

So what has gone wrong?

It may be that nothing really has gone wrong. It could just be that we have not yet had time to catch up with what is still a very

*The obvious exceptions, of course, are the computer and information processing industries themselves. According to the August 18, 1997, issue of the magazine *Business Week*, during 1996, the production of computers and semiconductors accounted for 45 percent of all U.S. industrial growth. The 1996 market value of Silicon Valley alone (so that excludes the two megagiants Microsoft in Washington and IBM in New York) was $452 billion. By comparison, Wall Street had a market value of $405 billion and auto giant Detroit a "mere" $113 billion. But one might reasonably—naively?—have assumed that the products the computer companies generate would have benefited their customers as well as themselves.

new technology—moreover, a technology that is still evolving, and which, when compared to other technologies, is still highly unreliable. (To be fair to the computer engineers, when you take into account the sheer complexity of the hardware and the software in use today, the remarkable thing is that it ever works at all.) It takes time for a major new technology to produce a major effect. For example, electric motors did not boost productivity growth appreciably until more than forty years after Edison installed the first dynamo in 1881, and it wasn't until 1919 that half of American plants were wired for power. It was later still before factories had reorganized to fully exploit the new technology. (Part of the "catching up" with the technology may well be that we need to change the way we measure productivity, of course. But as the examples just cited indicate, this is probably not the whole story.)

Another possible reason for the productivity paradox is that, despite the rapid growth, computer hardware still only accounts for between 2 and 5 percent of capital stock in most U.S. industries. Add in the cost of programs, telecommunications, and other related equipment and the total comes to about 12 percent. By comparison, it was only when they accounted for 12 percent of capital stock that railroads energized the economy.

A third possible reason is the issue I address in this book, that we do not yet understand how to manage information.

## MANAGING THE COMPANY'S MOST IMPORTANT ASSET

A modern company has many assets that have to be properly managed: physical plant, personnel, and the company's financial assets, to name just three.

To manage those different assets you need different kinds of expertise. If you are a manager, you may have the appropriate expertise within your company, or you may have to outsource.

For instance, large organizations generally have their own human resources department, where trained professionals oversee the hiring, training, and welfare of the work force. On the other hand, most companies will outsource their advertising.

In today's commercial environment, information is another asset that requires proper management. In many industries, information is now *the* single most valuable asset. Among the so-called knowledge-based economies, such as the United States or Western Europe, the management of information is taking an increasingly larger share of the cost of doing business, and that trend is likely to continue. Companies that are not able to manage their information efficiently will simply not be successful.

## THE GLUE THAT BINDS THE ORGANIZATION

It is a familiar cliché that information is the glue that holds together most of today's organizations. Unfortunately, this metaphor more often applies in a negative fashion. In many cases, information acts as the glue that holds the organization motionless when it should be the oil that keeps the wheels turning. How many times have you heard about a company that introduces a new computer system to improve its information management only to discover that, far from making things better and more efficient, the new system leads to scores of new problems that had never arisen with the old way of doing things? The new system is capable of providing vastly more information than was ever available before, but it is hard, if not impossible, to extract that information, or it's the wrong kind, or it's presented in the wrong form, at the wrong time, or it's delivered to the wrong person. Or there is simply too much of it for anyone to be able to use. What used to be a simple request over the phone becomes a lengthy battle with a seemingly uncooperative computer system, taking hours or even days, and eventually drawing in a whole host of people.

Why does this happen? The answer is that, despite all that we hear about living in the Information Age, what we are really living in is an age of information *technology,* or more precisely a collection of information technolo*gies.* We do not yet have an established *science* of information. As a result, we do not yet have the ability to properly design or manage the information flow that our technologies make possible.

In fact, many companies are not even aware that they need such a skill. Faced with the persuasive marketing of ever-more-powerful information systems, there is a great temptation to go for the "technological fix": If the present information system is causing problems, get a bigger, faster ("better") system. This approach is like saying that the key to Los Angeles's traffic problem is to build even more, and bigger, roads.

So what is the solution?

Just as a company has experts to manage its various assets, so too it needs experts to manage its information assets. Alongside the lawyers who handle and advise on contracts and the accountants who handle and offer advice on balance sheets, you also need "information scientists" who take care of the information assets.

Why aren't there "information scientists" at most corporations? The problem, as I noted a moment ago, is that there is as yet no "science of information" and consequently there are, at present, no "information scientists."* The world of information flow does not yet have the equivalent of a lawyer or an accountant. (Many might think this a good thing, but imagine what it would be like to be taken to court and have no recourse to a lawyer.)

## THE NEW IRON AGE

Why do we have so much difficulty saying what information is? One reason is that, although we can store information using various kinds of physical media, information itself is not physical; it is abstract. Physical objects store information in the same way that Cheshire Cats have grins. When the Cheshire Cat vanished, all Alice was left with was the grin; similarly, take away the physical representation of some piece of information and all you are left with is the information.

Whereas information is not physical, it is not purely mental either. Our thoughts are locked inside our head, but information is in some sense "out there" in the world. Whatever it is, information exists somewhere in between the physical world around us and the mental world of human thoughts. It occupies what I call "the information level." (See Figure 2-1 on the following page.)

In terms of our scientific knowledge, today's Information Age is reminiscent of an earlier era: the Iron Age. Imagine yourself suddenly transported back in time to the Iron Age. You meet a local ironsmith and you ask him, "What is iron?" How will he answer? Most likely, he will show you various implements he has made and tell you that each of those was iron. But this isn't the answer you want. What you want to know, you say, is just what it is that makes iron *iron* and not some other substance.

---

*Actually, there is an academic discipline called "information science," but this is not what I mean here. That kind of information science studies ways of organizing and retrieving information. It used to be called "library science." The name was changed to try to eradicate the stuffy image of elderly ladies staffing the library checkout desk on a voluntary basis.

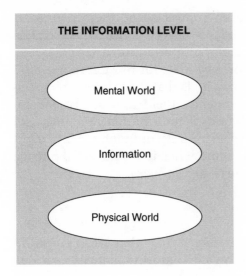

**THE INFORMATION LEVEL**

Mental World

Information

Physical World

**Figure 2-1** The information level.

For all that he may be a first-class ironsmith, your Iron Age Man cannot provide you with the kind of answer you are looking for. The reason is that he has no frame of reference within which he can even understand your question, let alone give an answer. To provide the kind of answer that would satisfy you, he would need to know about the atomic structure of matter—for surely the only way to give a precise definition of iron is to specify its atomic structure.

Today, in the Information Age, we are struggling to understand information. We are in the same position as Iron Age Man trying to understand iron. There is this stuff called information, and we have become extremely skilled at acquiring and processing it. But we are unable to say exactly what it is because we don't have an underlying scientific theory upon which to base an acceptable definition.

Incidentally, in addition to information science, which was mentioned earlier in the footnote on page 23, there is another field called "information theory." This is a branch of engineering mathematics developed over the past fifty years that investigates the amount of information that may be transmitted via a particular communication channel. The focus of attention is not on information as such but on signals: How complex a signal can be transmitted through a particular channel? However, as we shall see in the ensuing chapters, there is no fixed connection between a particular signal or configuration of objects and the information it represents. It all depends on the way the information is encoded. A complex signal might not represent any information. On the other hand, the transmission of a single pulse along a wire—1 bit—could represent an enormous amount of information. As a result, information theory is extremely useful for the engineers who design telephone and computer networks. But it is of little relevance if you want to study the way people communicate with one another.

What we need then, is a genuine science of *information*, a science on which we can build an understanding of communication. Situation theory, introduced by Barwise and Perry in the early 1980s and subsequently developed at Stanford University's Center for the Study of Language and Information (CSLI), was developed to provide the mathematical foundations of such a science.*

Situation theory allows us to distinguish between information and its representations and to examine the mechanisms whereby representations encode the information they do. That particular application was in fact one of the driving forces behind the development of situation theory. It makes an excellent point at which to start our own investigation.

## SUMMARY

Today, we have an information technology but not an information science. There is no generally accepted scientific description of what information is. As a result, we are often unable to manage information efficiently. Computer systems often cause as many problems as they solve. This may explain, in part, the productivity paradox.

The first step toward better information management is to understand what information is, how it arises, and how it is transmitted.

In the early 1980s, a group of researchers at CSLI developed a new mathematical theory—situation theory—to provide a firm theoretic framework on which to base such understanding.

As outlined in the Prologue, situation theory has already chalked up a number of successes.

In particular, situation theory provides a framework for studying the way information is represented.

---

*Some successful applications of situation theory were listed at the end of the Prologue.

# 3

# No Information Without Representation

## Why libraries don't contain information

Before you can learn how to properly manage information, you need to know what it is and what it isn't. By this I don't mean the kind of deep, precise knowledge of a science. As I pointed out earlier, we do not yet have such a science. But you do need to be clear about what is normally meant by the word "information" (people are notoriously sloppy about this), about how it is created, how it is stored, how it is transmitted, how it is processed, and how it is used.

One of the most common misunderstandings is to confuse information with its representation—words on paper, diagrams in books, bits on disks, or whatever.

For example, while we often speak of libraries as "repositories of information," that is not strictly true. What you find in libraries are books, not information. And those books don't contain information either, they contain pages. Similarly, those pages carry various markings, but no information. How does information come into the picture? By being *encoded* or *represented* by the markings on the pages. However, being encoded or represented is not the same as being "contained in."

This may all sound like we're splitting hairs, but appreciating the distinction between information and its representation is the key to efficient information management. (Situation theory was developed largely in order to analyze the way information is represented.)

To see just how important this distinction can be, look at a popular, if fantastical, example of how they can be confused. In the movie *Jurassic Park,* scientists recreate dinosaurs by obtaining dinosaur DNA from the dried blood found in mosquitoes sealed in amber. Scientifically speaking, this is not a highly realistic scenario, since, while insects dating from the dinosaur era have been found intact in amber, experts doubt that any dinosaur DNA could survive to the present day. But let's grant the movie makers that possibility. What then? Could scientists use that DNA to create a dinosaur?

Certainly, the DNA is the blueprint for a real-life dinosaur—it encodes all the information needed for a dinosaur to form. But the key word here is "encodes." The DNA is not itself the information required to create a dinosaur; it is a representation of that information. To recreate a living dinosaur, the required information needs to be recovered from that representation. For that, you need a "dinosaur DNA reader." In other words, you need a dinosaur egg! The chemicals in a dinosaur egg were able to "read" the information in the dinosaur DNA in order to create a baby dinosaur. The only hope of creating a living dinosaur today would be to discover a living egg, perhaps frozen deep in some Antarctic ice flow. DNA on its own would not be enough. Having a sample of dinosaur DNA is like having a computer disk without the right kind of computer to read it.*

A good example of the way environment affects the information obtained from DNA is provided by the *hox* genes, regulatory genes that help establish major body structures. In different animals, these genes can establish very different structures. In the fruit fly embryo, for instance, a *hox* gene called *Abd-B* helps define the posterior end of the embryo, while a similar family of genes in chicks helps to partition a developing wing into three segments.

---

*All right, maybe, just maybe, you could get a dinosaur by inserting dinosaur DNA into an egg of a prehistoric creature sufficiently like a dinosaur that has survived to this day, such as a crocodile. It's highly unlikely, but the only way to find out for sure would be to try it. Despite the low likelihood of success, this might be enough of a window for a Hollywood film director. In any event, the point to be aware of is that the DNA on its own is not enough; you need the right environment in order to extract the information from that DNA.

## SUPERCOMPUTERS AND KNOTTED HANDKERCHIEFS

Though we generally think of information as being stored in books and computer databases, any physical object may store information. In fact, during the course of a normal day, we acquire information from a variety of physical objects and from the environment.

For example, if we see dark clouds in the sky, we may take an umbrella as we leave for work, the state of the sky having provided us with the information that it might rain. On Halloween night in North America, a light on in the porch provides the information that it is acceptable for children to approach the house and ask for candy; no light indicates that the householders do not want to be disturbed. In rural parts of North America, setting the flag on the mailbox in the upright position indicates to the mail carrier that there is outgoing mail to pick up.

Animals too can acquire and act upon information. Dogs come running to us when called, cats respond to the sound of the cat-food can being opened, and many animals are aware that smoke indicates fire and that fire represents danger. It is arguable that plants and certain physical devices also acquire information and react accordingly. For example, a flower opens when it detects the sun's rays in the morning, and a thermostat will switch on the heating when it detects a drop in temperature.

These few examples indicate one significant factor concerning information that we need to take into account: Different people, animals, and perhaps plants and certain physical devices can obtain different information from the same source. A person can pick up a great deal of information about the surrounding air—how clean it is, the presence of any smells, whether it is warm or cool, how humid it is, and so on. A simple thermostat, on the other hand, can only pick up one piece of information from the surrounding air, the information whether the temperature is above or below the value set.

Going in the other direction, different objects or configurations of objects can represent the same information. For example, dark clouds in the sky, the reading on a barometer, and the weather report in the newspaper can all represent the information that it is likely to rain.

How can an object or a collection of objects encode or represent information? How can a part of the environment encode or represent information? And would all of this encoded and represented information exist if no people or other information-sensitive creatures were around to perceive it?

For instance, how does smoke provide information that there is fire, and how do dark clouds provide information that it is likely to rain? Part of the explanation is that this is the way the world is: There is a systematic regularity between the existence of smoke and the existence of fire and a systematic regularity between dark clouds in the sky and rain. Human beings and other creatures that are able to recognize those systematic regularities can use them in order to extract information. The person who sees dark clouds can take an umbrella to work, and the animal that sees smoke on the horizon can take flight.

Notice that we are definitely talking about *information* in these examples, not what the information is about. For example, people or animals that see smoke do not necessarily see fire, but they nevertheless acquire the *information* that there is a fire. And the sight of dark clouds can provide the *information* that rain is on the way long before the first drop falls.

In general then, one way information can arise is by virtue of systematic regularities in the world. People (and certain animals) learn to recognize those regularities, either consciously or subconsciously, possibly as a result of repeated exposure to them. They may then utilize those regularities to obtain information from aspects of their environment.

What about the acquisition of information from books, newspapers, radio, and other media, or from being spoken to by fellow humans? This too depends on systematic regularities. In this case, however, those regularities are not natural in origin like dark clouds and rain or smoke and fire. Rather they depend on human-made regularities, the regularities of human language.

In order to acquire information from the words and sentences of English, you have to understand English—you need to know the meanings of the English words and you need a working knowledge of the rules of English grammar. In addition, in the case of written English, you need to know how to read—you need to know the conventions whereby certain sequences of symbols denote certain words. Those conventions of word meaning, grammar, and symbol representation are just that: conventions. Different countries have different conventions: different rules of grammar, different words for the same thing, different alphabets, even different directions of reading—left to right, right to left, top to bottom, or bottom to top.

The linguistic conventions for any one country or population have evolved over thousands of years. In many ways, those conventions are quite arbitrary and liable to further change. What makes the entire system work is that the conventions of language are, at any one period, regular and systematic. The symbol sequence *cat* means the same thing to people all over the English-speaking world, it means the same thing today as it did yesterday, and it will mean the same thing tomorrow.

Of course, the systematic regularities in language are not natural regularities as are smoke and fire or dark clouds and rain. Rather they are social regularities, determined by a linguistic community. As such, they may be used by anyone in that linguistic community.

At an even more local level, there are the conventional information encoding devices that communities establish on an ad hoc basis. For example, a school may designate a bell ring as providing the information that the class should end, or a factory may use a whistle to signal that the shift is over.

There are even one-person information storage devices, such as the knot in one corner of your handkerchief, that today is supposed to remind you to pick up the laundry on the way home from work but which yesterday encoded the information that your wedding anniversary was coming up and you should remember to book a table at your favorite restaurant.

The fact is, anything can be used to store information. All it takes to store information by means of some object—or more generally, a configuration of objects—is a convention that such a configuration represents that information. For information stored by people, the conventions that may be used to store information range from those adopted by an entire nation (such as languages) to a convention adopted by a single person (such as a knotted handkerchief). In the case of nonhuman information storage, nature evolves its own conventions, such as the principles by which DNA encodes the information required to create a life-form (in an appropriate environment).

## THE HIDDEN COST OF INFORMATION

The thing to remember about information is that to recover information from a stored representation, you need to know the convention—the rules of the encoding. To calculate the true cost of any information, you need to add together the cost of obtaining or storing the representation and the cost of implementing the procedure necessary to recover that information from the representation.

The latter component is often overlooked. But in many cases, it is by far the greater cost.

For example, suppose you have just purchased a new Japanese computer. You open the instruction manual to discover it is written in Japanese. On one level, you could be said to "have" all the information required to operate the machine. But unless you understand Japanese, that information is not going to do you any good. In fact, for all practical purposes, you *don't* have the information.* What the manual gives you is a *representation* of that information. To recover the information from that representation, you need to be able to read Japanese. If you do not, then the cost to you of getting that information is far more than the cost of the representation—of the manual. If that particular computer only comes with a Japanese manual, then you have to add on the cost of learning Japanese or of hiring a Japanese speaker. If the computer is a million-dollar supercomputer that cannot be obtained from any other source, that additional cost may acceptable. But if you were buying a two-thousand-dollar desktop machine, you would probably decide to send back the Japanese product and buy a U.S.-made Dell computer instead.

To emphasize the crucial role played by the encoding scheme in the storage and transmission of information, I'll state it in the form of an equation—what I call the *information equation:*

> Information = Representation +
> Procedure for encoding/decoding

The information equation gives rise to an important cost version:

> Cost of information = Cost of representation +
> Cost of procedure for encoding/decoding

In the example of the Japanese computer, the decoding cost incurred in reading the manual was to the consumer's disadvantage. But the encoding/decoding procedures required to store and transmit information can work to your advantage. Here's how.

Computers and "information systems" deal exclusively in *representations* of information—digital representations in the case of the modern computer. Whether the symbol manipulations that a computer performs correspond to efficient or inefficient processing of

---

*In fact, the information management techniques described in this book, being motivated by highly practical purposes, will tell you that you *really* don't have the information—period.

*information* depends to a very large extent on the way the symbols represent the information. By coding a large amount of information using a comparatively small number of symbols, it is possible to get a significant gain in efficiency. The cost is shifted from the manipulation of the encoded information to the encoding/decoding process.

For example, when the cost of computer processing fell well below that of installing and maintaining telephone wires, it became more cost-effective for the telephone companies to encode, compress, and pack together messages prior to transmission and then unpack, decompress, and decode them at their destination rather than keep adding more line capacity. In that case, a highly complex encoding/decoding procedure allowed a large amount of information to be encoded in a relatively compact representation.

At this point it will be helpful to introduce a technical term that will be explained more fully later (see Chapter 6). "Constraint" is the term situation theorists use to refer to the regularities and conventions that enable some configuration of objects to represent or store information. This is not the most ideal choice of word, since we often use the word "constraint" to refer to some kind of restriction. But as is so often the case in science, the technical usage of the term just crept into the discussions in the early days, and by the time everyone realized that it might cause problems for newcomers to the field, it was too late.

In terms of constraints, the information equation, introduced above, can be written

$$\text{Information} = \text{Representation} + \text{Constraint}$$

Using the concept of a constraint, we can now take a closer look at the distinction between information and data that I mentioned earlier.

## INFORMATION OR DATA?

A common answer to the question, "What is information?," is to say that information is data that has meaning. This definition can be encapsulated in the equation

$$\text{Information} = \text{Data} + \text{Meaning}$$

This is a reasonable way to approach information, but to make any use of it as a *definition* you have to define what the two terms on the right mean: What is data and what is meaning? Unfortunately, it is not all that easy to find a definition for either term. It's a little easier if we start with the equation

Information = Representation + Constraint

since it is easy to say what constitutes a representation: Anything can be a representation! All that remains is to provide a definition of the term "constraint," a task we'll take up in due course.

In most cases, when people use the data + meaning view of information, they are, I think, assuming some collection of constraints, and then "data" simply refers to the representations of information via those constraints. For example, a spreadsheet full of figures would be data, since the figures (that is, the numerical marks on the paper or the computer screen) represent numbers (mathematical objects) and the structure of the spreadsheet enables those numbers to tell us something about the world—to give us information. Conceptually, this approach is fine in terms of trying to wrap your mind around the issue of what is information, but as I observed a moment ago, it does not lead easily to any kind of definition in simpler terms.

Another convention that some people use to try to get a hold on information is to distinguish between "Information", written with a capital *I*, and "information", written with a lowercase *i*. In this case, "Information" (with a capital *I*) corresponds to what I am calling simply *information,* and "information" (with a lowercase *i*) corresponds to my *data.* To emphasize the distinction and avoid confusion when speaking (and sometimes in writing as well), Information is generally referred to as "big-I Information," and information as "little-i information."

The relation between big-I Information and little-i information is indicated by the equation

Big-I Information = Little-i information + Meaning

People who study information technology in the (people-populated) workplace often use the big-I, little-i terminology. The intention is to use the two ways of writing the word "information" to reflect the two different meanings people attach to the word. Little-i information is what computers and other nonliving information processors process or manipulate. As a result, when computer professionals talk about "information," what they generally mean is little-i information. Likewise, little-i information is what you find discussed in books on "information technology." But, ultimately, what people want is big-I Information. When computer *users* talk about "information," what they usually mean is big-I Information. The two concepts of information come into contact when people want to obtain big-I Information from the little-i information that the computers provide. (Actually, this is not just a computer issue. People also obtain big-I Information from the little-i information that is provided by other people in the form of words, either spoken or written. We'll come back later to the nature of communication, both person-to-person and person-with-machine.)

The big-I, little-i distinction is useful when we want to discuss the design and use of information technology. But it does not provide an ideal basis for a scientific analysis of information (that is, of big-I Information), since the little-i information of this approach seems more or less the same as the data of the information = data + meaning approach discussed earlier—and as we already observed, that approach does not lead to a useful *definition* of information. Accordingly, in this book, I shall not use either the data + meaning or the big-I, little-i terminology. Instead, I will adopt the approach

Information = Representation + Constraint

As we noted earlier, since anything can be a representation, the scientific analysis of information can then begin with an examination of constraints.

---

## SUMMARY

The first step toward understanding information is to recognize the distinction between information and its representation.

Almost anything can encode information. The key to being able to encode information or to extract information from an encoding is to know the encoding scheme. This is captured by the information equation:

Information = Representation +
Procedure for encoding/decoding

Our twentieth-century information processing technology gives us devices that can store and process *representations* of information. Their efficiency depends not only on the speed and manner with which they process the representation but also on the efficiency of the encoding scheme.

Using the technical term "constraint" to refer to the regularities and conventions that enable some configuration of objects to represent or store information, we can write the information equation as:

Information = Representation + Constraint

This raises the possibility of analyzing information by starting with an examination of the concept of a constraint.

# 4

# The Dinosaur's Egg

------

## A NEW SITUATION

As we saw in the last chapter, for nature (or would-be theme park entrepreneurs) to recover from DNA the information required to construct a dinosaur, it takes a suitable "dinosaur-DNA decoder." Until some time around 65 million years ago (when the dinosaurs became extinct), nature itself provided just such a device: namely, the dinosaur egg. The dinosaur egg contained exactly the right mix of chemicals to "read" the information stored in the dinosaur DNA and use it to create a baby dinosaur. A similar process takes place in the mother's womb when a baby human is created. It's the egg and the womb that provide the environment needed to recover the necessary information from the representation. When we say that dinosaur or human DNA encodes all the information needed to construct another member of the species, what we mean—or what we *should* mean—is that the DNA encodes such information *within the appropriate environment*.

In other words, information is context-dependent. Not just slightly so, but in a significant and essential way. In fact, the key to obtaining information is *always* to be found in the context, not in the representation. There is nothing special about an object that encodes information—you can use almost anything. Information is not something intrinsic to an object; it is something *ascribed* to it by some form of information processor.

With that one observation, in our quest to understand information, we suddenly find ourselves thrust into a very different world. The emphasis shifts at once from texts, lists, tables, diagrams, charts, computer disks, DNA molecules, and knotted handkerchiefs—the things we might have expected to study—to the information processing devices (often people, but not always so) that use them. We have to transfer our focus from the dinosaur's DNA to the dinosaur's egg, and in particular to the all-important chemical soup that fills the egg.

The problem is that the sheer diversity of possible informational environments—of contexts for information—threatens to overwhelm us before our analysis has even got off the ground. The only hope for progress is to find a suitable level of generality at which to begin, and try to avoid getting bogged down in detail. This is exactly the situation in which two Stanford University researchers, Jon Barwise and John Perry, found themselves in the late 1970s and early 1980s, when they began their attempt to develop a new science of information.

Barwise and Perry decided that the only way to make progress was to think about environments, or contexts, *in the most general possible way*. Forget all their particulars, they said. Just work with "environments" in a very general sense. But what does this mean exactly?

Well, we all "know" what an environment is, right? It's a very general concept that includes physical environments, intellectual environments, cultural environments, manufacturing environments, political environments, economic environments, educational environments, and so on.

Some environments have physical aspects, others are highly abstract. In the case of a physical environment, we can be physically *in* the environment, whereas being "in" one of the more abstract kinds of environments (such as a political environment) generally means that it influences and conditions some or all of the actions we perform.

All right, said Barwise and Perry, take this general, intuitive concept of an "environment" and try to build onto it a framework for understanding information—try to describe information *relative* to our naive understanding of environments.

One problem with thinking in terms of environments—of even using the word "environment"—is that it immediately raises the questions "Environment of what?" and "Environment for what?" In order to get away from those questions and to start at the simplest possible level, Barwise introduced the term "situation" to refer to any possible environment or context (of and for anything whatsoever).

Of course, the word "situation" also carries with it an everyday meaning, but that everyday meaning is broad enough to encompass anything that might be termed an environment or context, and it includes a lot more besides: empty rooms, historical situations, imaginary situations, sporting events, and so on, none of which would normally be termed an environment, though any situation *can* be an environment, of course.

It was this concept of a (very general) *situation* that Barwise and Perry took as the starting point of their new theory of information. They called it, naturally enough, situation theory.

In situation theory, the aim is not to analyze the concept of a situation as such. Rather, situations are the starting point, and we rely on our intuitive understanding of what the word "situation" means. We then investigate the way situations give rise to encodings and decodings of information. This may appear to be cheating, but in fact the approach is not uncommon in science. For example, in the early days of the atomic theory of matter, physicists simply postulated the existence of tiny particles of matter, which they called "atoms," and then examined the way atoms could be combined together to give the variety of forms of matter we find in the world.

Of course, as we well know, physicists were subsequently able to analyze the nature of atoms, and the same has started to happen with situations. But to get our study of information off the ground, we can do here exactly what Barwise and Perry did and leave the concept of a situation essentially unanalyzed, just as physicists did in the early days of atomic theory. (Incidentally, the same thing happened again with the idea of a gene. At first, genes were simply postulated as the "things" that carry the information to influence hereditary inheritance. Subsequent research then showed that genes are parts of DNA molecules.)

From the point of view of science, then, situations were a new kind of entity to be studied. Situation theory examined the way that situations give rise to the creation, storage, and transmission of information. It turned out to be a very useful way to think about information, which has already led to the advances mentioned at the end of the Prologue.

To those of us working on the early development of situation theory, situations seemed at first to be a very strange kind of entity on which to build a "physics of information," a new science of information. It seemed hard to grasp the fact that something as nebulous as, say, the current situation in the Middle East could be part of a rigorous, mathematically based, scientific theory. (What are the geographic limits of the situation in the Middle East? Which people are constituents of this situation and which are not? When did it start? Does it remain the same from day to day?) But as we started to work with the new theory and became more familiar with it, we began to realize that situations are as natural and unavoidable entities to consider in the study of information and communication as are atoms in the study of matter. In other words, situation theory is indeed "the physics of information and communication."

## WHY DO YOU STOP WHEN THE TRAFFIC LIGHT TURNS RED?

To try to get a sense of how situation theory helps us to understand information flow, consider the following familiar scenario. You are in your car, and you have just entered an unfamiliar city— one you have never before set foot (or car) in. You come to an intersection controlled by traffic lights. As you approach the intersection, the lights turn red. What do you do? You stop. Why?

Saying "Because the light is red" or "Because a red light means stop" won't do. I'll just ask "Why?" again. What I want is a situation-theoretic explanation of the mechanism involved.

First of all, the traffic light provides a means to represent information. When the light is red, the information represented is that motorists should not proceed; when it is green, motorists should proceed. But what plays the role of the dinosaur's egg? What is required in order for you, the motorist, to obtain that information from the light?

Well, your arrival at the intersection constitutes a situation, let's call it s. The situation s has physical location and a brief duration in time. You, your car, and the traffic light are all constituents of that situation. (So too are a lot of other things—maybe there is a snail trying to cross the road—but they will not be germane to our analysis.)

At this stage it is tempting to say that s is the dinosaur's egg. But that can't be the whole story. You have never before been in this city. The situation s in which you find yourself is brand new to you. Most likely the *only* features of s that are familiar to you are that s has an intersection in it with a traffic light that is showing red.

This is pretty minimal information about *s*. Just think about all the hundreds of other things that one might be able to say about the situation *s*, given enough time to investigate! Yet this is all you need! In fact, all you need to know is that you are facing a red light; it doesn't matter whether there is an intersection or not.

Being able to drive involves knowing how to respond in a whole range of different kinds of situations. In particular, it involves knowing how to respond *in any situation* in which there is a traffic light showing red. The key to your responding appropriately to the particular red light you encounter in the novel situation *s* is that you know how to act in *any* situation where you face a red light. It is the fact that all motorists should behave the same way, on *any* occasion when they find themselves in a situation of a certain type, that makes the particular red light in *s* carry the information that you should stop.

It is that word "any" in the above paragraph that is the key to the red light being able to provide you with the information that you should stop. If red lights sometimes meant stop and sometimes meant go, you would not know how to behave in situation *s*. The particular red light in *s* would not convey the information that you should stop.

Though the familiarity and extreme simplicity of this particular example may lead you to think we are going round in circles, we have in fact made a key, and fundamental, observation about how information can arise and how it can be encoded and transmitted: An object in the world can represent information by virtue of that object being in a situation *of a certain type*.

When you say you know how to behave when faced with a red light, what you mean is that you know how to behave in any specific situation of a particular type.

What holds for red traffic lights also holds in general: Information arises by virtue of situations being of certain types.

To take the analysis a stage further (and thus completely identify our dinosaur's egg for the traffic light), we need to understand what it means for a situation to be of a certain type.

## WHY YOU DON'T WEAR SLIPPERS TO WORK

The chances are, you don't wear slippers to work. Why is that? Well, on any particular occasion when you are at work, call that situation *s*, it is not appropriate to wear slippers. On the other hand, in a typical home situation $h$, it is appropriate for you to wear slippers, and perhaps you do.

How do I know this about you? After all, you and I have probably never met, and there are many distinct work situations—one for each reader of this book—and in practically every one of them, it is inappropriate to wear slippers. Likewise, there are many distinct home situations, and in practically every one of those, it is appropriate to wear slippers. So it must be the case that all those individual work situations have something in common, and likewise all those individual home situations have something in common. Indeed, those common features are reflected in the very language that we use: "work situations" and "home situations." Now, we may find it difficult saying exactly what those common features are, but in practice we have no difficulty knowing whether we are in a work situation or a home situation—and thus in knowing whether slippers are inappropriate footwear or not. It is by virtue of our being able to recognize these common features that we generally avoid violating the associated patterns of footwear behavior.

In situation-theoretic terminology, whenever we find ourselves faced with a range of situations that exhibit such a common feature, we say that all the situations sharing that feature are of the same *type*. Thus, there is a type of work situation, a type of home situation, a type of red traffic light situation, and so forth. Types are a means of classifying situations. As such, types are (like situations themselves) another of the concepts situation theory adopts from everyday life and uses in a technical (though very natural) way.

This may seem a bit mysterious, but it is no different from saying that people can be classified by race, or by profession, and so forth. Any American is of type "American," an Indian is of type "Indian," and so on. Sometimes we classify by finer types, speaking of "African American," "Native American," "Irish American," and so on, when we classify by racial origin, or "Californian," "New Yorker," "Mainer," and so on, when classifying by state of residence (or maybe of birth).

## EVERYONE IS YOUR TYPE

When you stop to think about it, it becomes clear that the ability to recognize types of things lies at the basis of much of human cognition and communication. Humans are type recognizers. So too, it appears, are various animals. For example, bees can recognize the types of certain flowers, and pet cats and dogs recognize the type of feeding bowls and the type of doorways.

Many of the words in our language refer to types: types of objects, types of actions, and so on. For example, nouns that denote things do so by making reference to types, not to the specific object or action itself: the word "house" refers to any house, not one particular house; the word "car" refers to any car, not one particular car; the word "walk" refers to any walking action by any person or legged animal; the word "run" refers to any running action; and so on. Such words can of course be used on any particular occasion to refer to a particular thing or action, but such reference is only possible because the meaning of the word refers to *types* of things or actions.

The recognition of types lies behind much of our ability to obtain information from our daily environment. For example, suppose you look up and see dark clouds in the sky, and you say to yourself, "It looks like it might rain today." On the basis of one situation, the cloudy sky right now, you infer information about another situation in the future, the weather in that region later in the day. The state of the clouds encodes information about the possible future weather. It does so because of a systematic relation between skies of a certain type and subsequent weather of a certain type, a relationship you are aware of. You know that when the sky situation is of the *type* "dark clouds," it is often the case that the weather (or environmental) situation is subsequently of the *type* "raining." That is what it takes for dark clouds to provide you with the information that it is likely to rain. On its own, the actual dark cloud formation that you see on one particular occasion would not, in itself, tell you anything. It is a one-off event. It is by virtue of the sky being of a recognizable type that you can obtain the information you do.

Types are fundamental to human life. If we were not able to recognize types, the world would always be presented to us anew, and we would be unable to acquire information from our environment or to make any reliable inferences based on prior knowledge or past experiences.

Many of the times we acquire information, it is by virtue of "naturally occurring" types that we have learned to recognize. This is true of dark clouds and rain, for example. These are naturally occurring types, and the relationship between them is natural. On other occasions, the types and our response to them are a result of human convention. In the case of the traffic light, the relationship between the type of a red light and the information that we should stop the car is human-made and enshrined in our legal code.

Now, at last we are homing in on the true dinosaur's egg for traffic lights. It is the relationship between two types: the type of situation in which a motorist is faced with a red light and the type of action that motorist should take (namely, stop the car). The environment in which the red light provides the information that the motorist should stop—the egg—is the legal framework of traffic laws.

This observation is hardly a surprise. But then we are not in the business of trying to make surprising discoveries. We are trying to understand the mechanisms that underlie a familiar, everyday activity, namely the acquisition, storage, and transmission of information.

So now we have the following picture: A situation $s$ encodes information by virtue of that situation $s$ being of a certain type. (A variant is that an object encodes information in a context situation $s$ by virtue of the object having a certain configuration and the situation $s$ being of a certain type.)

Incidentally, that last observation tells us that there was something we left out of our account of the role played by the dinosaur egg. Namely, it was by virtue of the dinosaur's egg *being an object of a certain type* that the DNA inside the egg encoded the information to make another dinosaur. Of course, that type was simply "being a dinosaur egg," so this is beginning to seem circular. That's often the problem with simple examples used for illustration. In order to bring ourselves back from the realm of abstract theory, in the next chapter we'll take a look at some real-life examples of the role played by context in the transmission of information.

### SUMMARY

We have identified two everyday concepts that play a significant role in the encoding and transmission of information: situations and types.

Since the representation of information is always relative to a context or environment, in order to understand information we must first understand environments.

Following Barwise, we use the word "situation" to refer to any kind of environment. Situation theory develops an understanding of information based on this initial (and initially unanalyzed) idea of a situation.

Human beings (and, to a greater or lesser extent, various other animals, plants, and human-made devices) are able to act "rationally" by being able to recognize *types* and to respond accordingly.

*Types* are the regularities or commonalties shared by objects or situations.

In general, an object or a situation represents information by being of a particular type.

# 5

# Into the Valley of Death

------------------------------------------------------------

## JUST TELL ME WHAT I NEED TO KNOW

Writing in his 1989 book *Ideas and Information*, Arno Penzias says, "As work becomes increasingly information intensive, organizational success will depend more and more on giving each individual contributor needed information at the right place at the right time and in the right form."

If we were talking about supplying water or electricity or sheet metal, or almost any commodity other than information, it would be a straightforward matter to follow Penzias's advice. Given adequate resources, the only thing that would prevent us from getting the requisite materials to the right people at the right time would be our own incompetence. But when it comes to information, things are far less clear-cut.

As we have seen, information is not a precisely defined thing. It is certainly not the same as markings on a piece of paper or bits on a computer disk. Those are just representations of information. Unfortunately, it is definitely the *information* that people require to do their jobs; the representations on their own are not enough.

You might think that with sufficiently well-trained people, it should be enough to provide them with the representations of the information they need. They could retrieve the information from those representations. If that were the case, then we could of course meet Penzias's demand: Just set up a distribution network for the representations that meets the requirements.

Unfortunately, such an approach is not going to work. Even in highly specialized working environments, with highly trained specialists, huge problems can arise with attempts to transmit information, problems that can lead to what is known as an *information bottleneck,* where the flow of information slows to a virtual standstill, holding up progress in various parts of the company.

To appreciate not only why information bottlenecks can arise but also why they are almost inevitable, just take a look at the oldest and most familiar kind of information transfer, where one person talks to another face to face. Though in practice we almost always overlook it, there are two distinct contextual situations involved in a conversation, one for each of the two people talking. For all that each may "understand" the words spoken by the other, the meaning each intends his or her words to convey depends on his or her particular context. The success of the communication depends on each person knowing enough about the other's context to recover the information from the words spoken.

The problem is that you can't encode the context within the message. The most you can do within the message is try to provide an indication of what that context should be. In the case of a conversation where there are no distractions and no great time pressure, the feedback mechanisms that form part of the ritual "dance" that is human conversation generally ensure that things don't go far off course. But under different circumstances, things can easily go wrong, and when they do, the result can sometimes be tragic. A famous case occurred in the nineteenth century, when the misinterpretation of a single word led to one of the most famous military disasters of all time: The Charge of the Light Brigade, immortalized by Alfred Tennyson in his epic poem of that title.

## THE VALLEY OF DEATH

In March 1854, following Russia's attempts to expand her empire, Britain and France had declared war on Russia. On September 14, British and French forces invaded the Crimea, a northern peninsula on the Black Sea that is today a part of the Ukraine. The aim was to capture Sebastopol and destroy the Russian fleet. Following a major victory at the Alma, when a 40,000-strong Russian force was defeated, the allies advanced toward

Sebastopol, putting it under siege on October 17. The British troops were led by Fitzroy Somerset, Lord Raglan.

On October 25, the Russians attempted to break the siege, launching an attack on the British positions near Balaclava, a small town 6 miles to the southeast of the city. The attack was repelled by the guns of the British Heavy Brigade. Raglan ordered the cavalry, his Light Brigade, to move in and sweep the enemy from their redoubts while they were still in disarray. His order to Lord Lucan, the commander of the Light Brigade, read:

> *Cavalry to advance and take advantage of any opportunity to recover the Heights. They will be supported by the infantry, which have been ordered to advance on two fronts.*

When Lucan received the order, he took it to mean that he should begin his advance only when the supporting infantry arrived. Seeing none, he waited. Meanwhile, the Russians began to recover their positions, and started to drag away captured British guns.

Forty-five minutes later, Raglan—who from his hilltop position could see what the Russians were doing—asked his aid Airey to send a second message. It read:

> *Lord Raglan wishes the cavalry to advance rapidly to the front—follow the enemy and try to prevent the enemy carrying away the guns. Troop Horse artillery may accompany. French cavalry is on your left. Immediate. Airey.*

Airey gave the note to his deputy, Captain Nolan, to deliver. As Nolan sped away on horseback, Raglan shouted after him, "Tell Lord Lucan the cavalry is to attack immediately."

Lucan read Airey's message. It made no sense. From his position down in the valley, he could see neither the enemy nor any guns, apart from the massive Russian force holding what looked like an invincible position at the far end of the valley. Seeing Lucan's indecision, Nolan passed on Raglan's verbal message: "The cavalry is to attack immediately."

The order seemed clear. Under Lucan's command, the 673 soldiers of the Light Brigade started to make their way down the valley toward the Russian position. In the carnage that followed, 272 of them were killed. The Light Brigade was destroyed. They never had a chance. What is more, they were never supposed to advance on such an impregnable position. From Raglan's elevated position on high ground, he looked down on the routed Russians, pulling back and taking British guns with them. That was the "front" he meant when he issued the order for the cavalry to "advance rapidly to the front." But the only enemy Lucan could see was the one at the end of the valley. That was his "front." And when, like a good soldier, he finally agreed to obey the order he had been given, that was the front he attacked. The valley along which he advanced would become Tennyson's "Valley of Death."

Their advance stalled, the British troops had to spend the entire winter holed up in the Crimea, and it was not until September of the next year that they finally took Sebastopol. The war ended the following February. Altogether a considerable cost to pay for misunderstanding the referent of the single word "front."

The problem was not that anyone failed to understand the word "front." Rather, the tragedy arose because that word had two different referents, depending on different contexts. If Lucan and Airey had been able to engage in a normal conversation, almost certainly the misunderstanding—and the ensuing tragedy—would have been avoided. But given the actual circumstances, with Lucan and Airey in different locations and Raglan and Nolan acting as intermediaries, to say nothing of everyone being in the heat of battle, it is not at all hard to understand how things went wrong.

On the other hand, things can also go wrong away from the battlefield, even when every effort is made to ensure that information flows smoothly and efficiently. The following is an example from the information technology industry.

## AN INFORMATION BOTTLENECK IN THE COMPUTER INDUSTRY

In 1989, I was approached by Duska Rosenberg, a computer scientist specializing in human factor issues in the information technology industry. She had been trying to figure out what exactly had gone wrong when a large British manufacturer and supplier of mainframe computer systems had tried to automate part of its own information system. She had carried out an extensive ethnographic study, involving many hours of interviews with everyone who had tried to use the system. She suspected that the problem was caused by the interaction of different contexts, but there were at the time no analytic tools with which she could verify her suspicion.

Since 1983, situation theory had been under development as a basis for a linguistic theory, and Rosenberg knew about it, but no one had tried to apply it to the kind of business data she had collected. Indeed, my initial response, when she approached me and asked if I would collaborate with her to use situation theory to analyze her data, was that the problem she was dealing with was far too complex for our fledgling theory. Fortunately, Rosenberg persevered, and eventually the two of us did collaborate, with great (and, on my part, initially unexpected) success.*

Essentially, what Rosenberg and I did was provide the data with an architecture or structure that allowed us to make sense of it. As such, our approach was similar to that of investigators who are called in after a major airline disaster. Investigators start with the masses of evidence available, the wreckage, the data from the flight recorders, the evidence of any witnesses, the aircraft's maintenance log, and so on, and they have to provide that data with an architecture—they have to piece it all together in a way that provides them with the information they require (in that case, what caused the crash).

Here, in brief, is what we did. The data Rosenberg collected focused on the company's repair activities. When a customer informed the company of a fault in their computer system, a field engineer was sent out to the site. At the same time, the customer service desk started to generate a simple document called a Problem Report Form (PRF). The PRF was a simple slot-and-filler document on which could be entered various reference numbers to identify the customer and the installed system, the fault as reported by the customer, the date of the report, the date of the engineer's visit, the repair action taken, and any components replaced.

The PRF was a valuable document, providing the company with an excellent way to track both the performance of their computer systems and field engineers as well as the demand for spare parts. In particular, by analyzing the data supplied by the forms, the company could identify and hopefully rectify the weakest components in the design of their systems.

Because of the highly constrained nature of the PRFs, the highly focused nature of the domain—computer fault reporting and repair—and the fact that the important technical information on the forms was all entered by trained computer engineers, the company had expected that the PRFs would form the basis of a highly efficient source of information for all parts of the company. But that

*The result of our joint work was published in Devlin & Rosenberg, 1996.

was not how it turned out. Both in the early days, when the PRFs were paper documents, and later when they were replaced by an electronic version, experts faced with reading the forms encountered great difficulty understanding exactly what had gone wrong with the customer's system and what the engineer had done to put it right.

The problem was magnified when the company tried to develop an expert system to handle the information provided by the PRFs. It was at that stage that Rosenberg—an expert in analyzing problems with expert systems—was brought in.

Her initial investigation led her to suspect that the problem was contextual. The initial entries on the PRF were made by the customer service personnel (in their context), who based some of what they wrote on what they were told by the customer (there's a second context—the customer's); subsequent entries were made by the field engineers (a third context). The PRFs were then read by personnel in various different sections of the company, each having different areas of expertise and different education and training backgrounds—in all some five or six different reading contexts.

With so many different contexts involved, Rosenberg was fairly sure that was where the communications problems lay—that was why people trying to read the PRFs complained that they often had difficulty getting from them the information they required.

I'll give just a couple of the simpler instances of how the different contexts gave rise to different ways to interpret a completed PRF. First, the form was intended to provide information on how a particular computer system fault was repaired. But even the word "fault" was highly context-dependent. Several of the many PRFs Rosenberg examined turned out to involve nothing more than paper jammed in a printer. For the customer, with the system inoperable, it was faulty. For the field engineer sent out to investigate, *there was no fault at all*—there was "just a paper jam," something that the on-site operator should have dealt with instead of calling the company to send out a highly trained field engineer.

Second, field engineers often found that the quickest way to get the customer's system back into operation was to replace the smallest easily replaceable unit to which the fault could be traced. From both the customer's and the field engineer's point of view, the "fault" was corrected. But when the replaced unit was sent back to the diagnostic lab, it was still faulty, and the lab engineers were still faced with finding out what was wrong. For some of the people reading the PRF, including the customer service personnel, the document told them what the fault had been. For others, including the quality control experts, the PRF did not tell them what had gone wrong. After all, the very fact that a PRF had been generated at all told them that, from the customer's point of view, *something* had gone wrong. Knowing that the fault had been in a printer and not a CPU was little better for the quality control people. They wanted to know exactly what individual component had failed.

The two instances just given are simple ones, which I chose for illustration purposes. They illustrate the way that context can affect even the interpretation of the word "fault." Though that word is admittedly the most fundamental term of all in the domain of fault repair, serious misunderstandings were unlikely to occur as a result, and in any case could easily be rectified. The same was not true for many of the other context problems in the PRFs, several of which Rosenberg and I list in our book.

Using situation theory, Rosenberg and I were able to carry out a detailed analysis of the way context affected the information conveyed by the PRFs. Our work led to both a redesign of the PRF itself and, more significantly, a restructuring of the procedures surrounding the completion and use of the documents. Taken together, both changes led to better information flow and improved efficiency in the company.

As Rosenberg and I showed, through use of analytic tools such as situation theory, it is possible to improve the flow of information. But such an analysis takes both time and money, and typically it is only carried out after a major problem has arisen. Are there quick and easy ways to eliminate or minimize the possibility of such misunderstandings before a problem arises?

## HOW TO AVOID CONTEXTUAL MISUNDERSTANDINGS

How can you make sure that costly misunderstandings do not arise? How can you avoid your own equivalent of the disastrous Charge of the Light Brigade? How can a company make sure it does not find itself faced with the kind of information bottleneck created by the PRFs?

The simple answer is, you can't. Whenever you have two or more people communicating, or one or more people communicating with one or more machines, you have two or more contexts in which information is represented and interpreted. And whenever you have two different informational contexts, you face the possibility of different interpretations.

What you can do is be aware that contextual misunderstanding is an ever-present possibility, and do what you can to minimize the likelihood that it occurs. In the case of communication that is *not* face-to-face, you can try to ensure that there are feedback possibilities built in to the communication network. For instance, when I try to deposit a large check (more than a few hundred dollars) into my bank account using the ATM machine, the machine queries the amount, not once, but twice, before it accepts the deposit. The system has clearly been set up to minimize the error caused by an incorrectly placed decimal point or a doubly entered digit, that would result, for example, in my entering $5000.00 when I really wanted to deposit $500.00.

Another example of feedback checking is with the superb Macintosh computer interface, which always asks for your confirmation before carrying out some instruction having serious consequences such as deleting a file or closing a file without first saving your most recent work on it.

In the case of a typical business meeting, there are two things you can do to minimize the possibility of misunderstanding. The first is to "recap" at periodic intervals. The second thing you can—and should—do is, immediately after the conversation is finished, summarize what has been agreed in writing and circulate it to everyone else who was involved in the meeting. Your summary does not have to be long. You are not writing minutes, where everything said is recorded; you are just trying to summarize what was agreed as a result of the meeting. For the same reason, make your note informal. Use too much formality and it begins to look as though you are trying to create a legal document to which they can be held accountable. After you have circulated your note, you may hear back from one or more of the meeting participants, either with confirmation or corrections. If not, and if the meeting was a small one, say, with no more than three or four participants, you may find it useful to contact the others a couple of days later and ask them if they agreed with your summary. Again, the thing to avoid is any suggestion that you are trying to create a binding legal document. Your aim is simply to ensure that the ever-present likelihood of contextual misunderstandings is reduced to an absolute minimum.

## SUMMARY

Different contexts can result in the same term being interpreted in different ways. In ordinary, real-time, face-to-face conversation, the feedback mechanisms that are part of all conversation usually avoid serious misunderstandings. With other forms of communication, misunderstandings happen easily and more frequently than is generally appreciated.

By designing working procedures carefully, contextually caused misunderstandings can be reduced. But they can never be eliminated, and as a result the best strategy is to always be aware that they can happen. This is especially true when the consequences of misunderstanding can be costly.

In important meetings, the possibility of misunderstanding can be reduced by means of frequent "recaps" of what has been agreed so far. Still more security can be obtained by circulating a written summary of what you think has been agreed to, immediately after the meeting has ended.

# 6

# Acting
# with Constraint

---

The last chapter illustrated the crucial role played by context in the transmission of information. Prior to that, in Chapters 3 and 4, we introduced some machinery to help us analyze that way context and representation combine to yield information. Specifically, in a suitable situation, an object or configuration of objects could represent information by virtue of a constraint, as indicated by the equation

Information = Representation + Constraint

We put off until later saying exactly what constitutes a constraint. Now it's later.

Once we have formulated a precise definition of constraints, we will have all the machinery required to start to develop a science of information.

We begin with language.

## THE LINGUISTIC TYPE

The bell rings and the pupil or the worker knows that the class or the work-shift is over. How do they know that? After all, each particular ring of the bell is a brand new event that has never happened before.

As we saw in Chapter 4, what makes the ring of the bell useful as an informational device is that each ring is a particular *type* of event, a type of event that the pupil or worker learns to associate with another type of event, the ending of the class or shift.

In fact, many places have a more extensive code for bell rings. For example, a single short ring might indicate the end of class, a double ring the end of morning break, and a long ring the end of the school day. In such cases the relevant type is not just that of "a bell ringing" but the particular kind (a particular *type*) of bell ring.

Language functions in a similar fashion. Words—other than grammatical words such as prepositions—are types that are connected to things in the world. For instance, many nouns are types that are systematically linked to types of objects and many verbs are types linked to types of actions. Since you may well not have thought about language in this way before, let's take a closer look at this idea. It may seem a bit convoluted, but the benefit is that it will prepare the way for us to finally come to grips with information.

We often think of words as sequences of letters, but that is not really accurate. For example, how many words are there on the following line?

Apple apple APPLE *apple APPLE* **apple** apple **APPLE** apple

You could probably try to make an argument that the answer is nine, but I think most people would say that there is just one word, the word "apple," written down nine times. This is certainly what the linguist would say. What you see on the line are nine distinct *presentations* (or *instances*) of the one word "apple." Each presentation—each sequence of five letters—is of the same *type,* and it is that type that we call the word "apple."

Now, we can use the word "apple" to refer to an actual apple since we are aware of a specific connection between the word type "apple" to a certain type of object in the world—the type that all apples have in common.

More generally, language serves to refer to things and events in the world, and to convey information about the world, by virtue of connections between types—between linguistic types (which we call "words") and types of things and events in the world. Clearly then, we need to take a look at those connections between types.

## THE HIDDEN THREADS

Consider the following examples of what we sometimes call "rational action."

The motorist who stops at the red light does so because she is aware of the law that connects the type of a red traffic light to the type of action where the motorist stops the car.

The pupil for whom the ringing bell indicates that class is over is aware of a connection between the type of a bell ring and the type where the class ends.

The English speaker who uses the word "apple" to refer to a particular apple is aware of (and utilizes) the link between the type we call the word "apple" and a particular type of object in the world.

The person who grabs for his umbrella on seeing dark clouds in the sky does so because he is aware of the systematic connection between that particular type of sky (that is, full of dark, foreboding clouds) and a particular type of weather that often follows, namely, raining.

The thermostat that switches on the heating when the temperature drops below 65 degrees Fahrenheit does so because it has been set so that whenever the environment is of a certain type, namely, temperature below 65 degrees, it performs an action of a certain type, namely switching on the heating.

In each of these examples, the behavior of the person or the device, on any particular occasion results from, or is guided by, a systematic link between two types. What are these links and how do they arise? When we answer this question, we will have the last piece we need in order to come to grips with information.

In the case of the motorist at the traffic light, that link is established by law. For the ringing bell, the link is established by local convention within the school. With language use, the links between words and the things in the world that those words refer to are an established part of what it is to know how to use a particular language. The connection between dark clouds and rain is a natural one—that's just the way the world is, and we come to recognize that connection through experience. Finally, the thermostat is *designed* to function in the way it does—its construction incorporates the connection between temperature and heating control.

In each case, "rational action" results from a link between types. Those links can arise in different ways: by legal fiat, by local convention, by the world simply being the way it is, by design, or whatever. What makes them useful in guiding action is that they are systematic and reliable—they hold all the time, or at least a goodly proportion of the time.

Situation theory has a special term for these links: *constraints*. As I remarked once already, this is strictly a technical use of the word "constraint." To all intents and purposes, we can take it to be unrelated to its everyday meaning of some kind of limit to our behavior.

Thus constraints can be (human-made) laws, rules, conventions, natural and physical laws, and so on.

Constraints are the hidden threads that connect information with its representation. (See Figure 6-1.) A knowledge or an awareness of the relevant constraint, or an adaptation to it, is what enables a person to acquire the information represented by way of that constraint. For example, familiarity with the constraint that smoke comes from fire enables a person or an animal to infer the presence of fire from the appearance of smoke.

The constraints that connect words of a language (and words are types, remember) to certain types of things in the world are what enable us to use language to represent information. More precisely, the constraints connecting words to types of things in the world are *part* of the encoding process. The linguistic and logical structure of the language—the grammar—also plays a role. The grammar can also be expressed as constraints. (For example, is "run" being used as a noun or a verb? The constraints of grammar tell us that.)

In most cases it is not necessary to know anything about the origin of the constraint in order to use it to obtain information. In fact, you don't need any real awareness of the constraint. After all, a

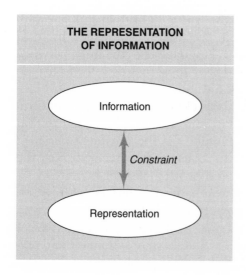

**Figure 6-1** The representation of information.

thermostat is not in any sense aware of how and why it operates the way it does. Nevertheless, it can process information well enough to maintain the room at a constant temperature. And most of us are not consciously aware of the constraints that constitute our working knowledge of our native language, though when we learn a second language as adults we generally find ourselves explicitly committing to memory the various constraints.

## THE RECIPE FOR INFORMATION

Now, at last, we have identified the three key ingredients that, when mixed together appropriately, give us information:

♦ situations

♦ types

♦ constraints

To complete the picture, we need to be a bit more precise about what we mean by information.

The thing to notice is that information is always information *about something*—information tells us something about something. This observation allows us to develop a theoretical approach to the concept of information using types: We shall assume that information always takes the form of a statement that *some object is of some type.*

If $a$ is any object and $T$ is any type, we shall use the abbreviation

$$a : T$$

to indicate that $a$ is of type $T$. Using this new notation, our assumption is that information is always of the form $a : T$ for some object $a$ and some type $T$.

A few moments spent thinking about this should convince you that we can always regard information in this way, even if, on occasion, it might seem somewhat unnatural.

For example, the information that John Smith is 38 years old is of the form $a : T$, where $a$ is John Smith and $T$ is the type "being 38 years old."

Again, the information that the profits of company $X$ are rising is of the form $X : R$, where $R$ is the type "profits are rising."

One more (we'll see many more examples during the remainder of this book): The information that it is raining is of the form $E : R$, where $E$ is the immediate environment (or the environment being referred to) and $R$ is the type "raining." As this example indicates, when we use ordinary language to provide information about a situation, we often omit explicit reference to the situation, and leave it to the context to supply the reference. We say "It is raining" rather than "It is raining in my environment."

Of course, the $a : T$ approach to information is only suitable for talking about single items of information. Often, when we use the word "information," we mean a whole mass of the stuff, such as a list, a database, or a spreadsheet. In those cases, the information is a collection of different items, each of the form $a : T$. To preserve the distinction between the two senses of the word "information"— the item sense and the collection-of-items sense—in my 1993 book *Logic and Information*, I coined the word "infon" to refer to a single item of information, something of the form $a : T$. (This is not completely accurate. In very technical work, "infon" has a slightly different meaning. But for our present purposes, that distinction is irrelevant.)

## HOW INFORMATION IS TRANSMITTED

Now that we have under our belts the concepts of situation, types, and constraints, we can describe the way that information is obtained and how it is transmitted.

Consider again the traffic light example from Chapter 4. You come to a red traffic light and you stop your car. Why? In informational terms, the red light provides you with the information that you should stop. But how does the red light provide you with this information?

As we observed, the significant thing about the traffic light that tells you to stop is that it is of a certain type that you recognize, namely the type of being red. As a qualified driver, familiar with traffic regulations, you are aware of a constraint that links the type of situation where a driver is faced with a red light and the type of situation where the driver should stop the car.

Figure 6-2 illustrates the scenario you are faced with. $C$ is the constraint that links the type ($S$) of situation where a motorist faces a red light and the type ($R$) of situation where the motorist stops the car. The actual situation you face is $s$. Because $s$ is of type $S$, in order to act as indicated by the constraint $C$ (that is, in order to remain compliant with the law), you should act so as to bring about

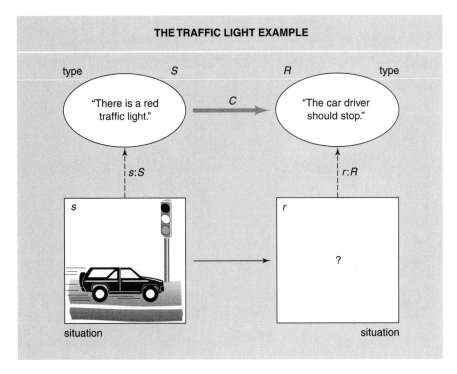

**Figure 6-2** The traffic light example: The scenario you are faced with.

a situation *r* that is of type *R*. That is to say, you should stop the car. Figure 6-3 on the following page illustrates the scenario that follows your stopping the car. The situation *r* that results is then of type *R*. Acting according to the constraint means bringing about a situation *r* that is of type *R*.

In the traffic light example, you, the motorist, have to act in order to bring about a situation *r* that fits the requirements displayed in Figure 6-3. For a slightly different example, where you do not have to take such action, suppose you observe a situation *s* in which there is smoke. Even though you don't see any flames, you nevertheless conclude, "There must be a fire." Why? Because the situation *s* that you observe provides you with the *information* that there must be fire. It does so because of the constraint that says "smoke implies fire," illustrated in Figure 6-4 on page 65. This constraint, denoted by *C* in Figure 6-4, links the type, *S*, of situation in which there is smoke with the type, *R*, of situation in which there is fire. In this case, the situation *s* that you observe will itself be the

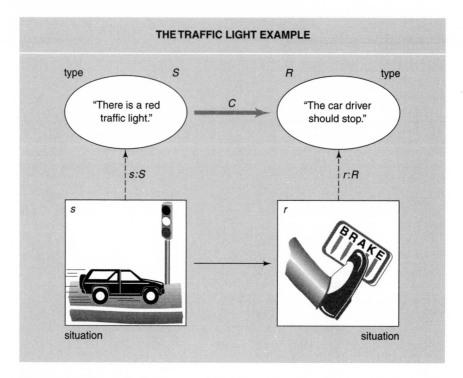

**Figure 6-3** The traffic light example: The scenario that follows (you stop the car).

situation where there is fire. You can see the smoke, so you know that the situation *s* is of type *S* (in symbols, *s* : *S*). The constraint *C* then tells you that the situation *s* must in fact be of type *R* (in symbols, *s* : *R*). Thus, even though you cannot see the flames, you can conclude that there is a fire.

Figure 6-5 illustrates the general framework that supports the flow of information. Given a constraint *C* that links the two situation types *S* and *T*, whenever there is a situation *s* of type *S*, then the situation *s* provides the information that some other situation *r* is of type *R*. The situation *r* may be one that you have to bring about, as in the traffic light example; or it may be the same situation, as in the smoke and fire example. Alternatively, the situation *r* might be *s* at some later time; for example, the constraint that says a dark sky indicates the likelihood of rain at that location later on. A fourth possibility is that *r* is a quite different location; for example, if you hear a loud bang (situation *s*) then you can conclude that there has been an explosion at some other location (situation *r*).

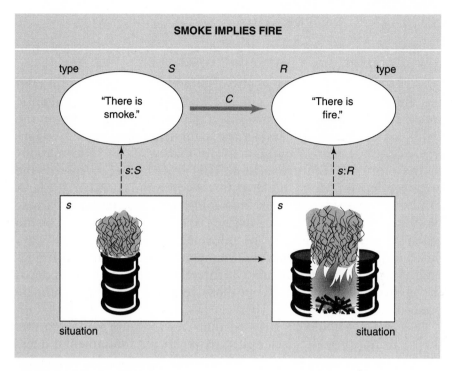

**Figure 6-4** Smoke implies fire.

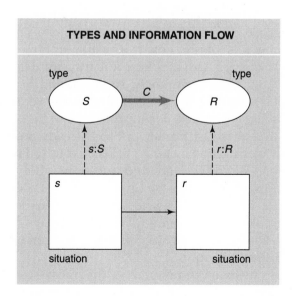

**Figure 6-5.** Types and information flow.

## SO WHAT IS INFORMATION?

Two things should be stressed about the approach to information developed so far.

First, the idea is purely to give a sound theoretical basis for talking about information—to put our understanding on a scientific footing. When it comes to using our theoretical framework "in the field," to design, improve, and manage information systems, we generally continue our normal practice of talking about "information" in an (informed) intuitive fashion. This is analogous to the way the engineer depends upon, but does not become embroiled in, the laws of physics when designing or managing the construction of a new building or other engineering project. Another analogy would be the manager whose decisions are informed by—and in many ways based on—sound economic and financial theory: The manager generally has neither the time nor the interest to become bogged down in the fine academic details, but those details support everything he or she does.

Second, it should be stressed that viewing information in this way does not provide a scientific answer to the fundamental question "What is information?" in the way that, for example, we can use the atomic theory of matter to answer the question "What is iron?" In terms of the "What is?" question, the Information Age is still very much at the same stage of ignorance as the Iron Age was for Iron Age Man.

What we have done is provide a framework to think about information. Using this framework, we can say how information arises, how it is stored, how it is acquired, how it is combined and otherwise processed, and how it is transmitted.

As the development so far should indicate, the framework we have derived is a very natural one—we derived the ideas of situations, types, and constraints by looking at some simple examples. Moreover, our framework has already proved itself to be extremely powerful. For instance, it has led to the resolution of a number of long-standing academic questions about language and information. (Some of those applications were enumerated at the end of the Prologue.) Another major application—the one that will be pursued in the remainder of this book—has been the provision of a basis on which we can develop more efficient methods to manage information.

It should be pointed out that most—though not all—of the observations and results that have so far been obtained using situation theory have confirmed what people already knew, or at least suspected. In that respect, situation theory has led to the rediscovery of a number of existing wheels. Far from being a surprise or amounting to a critique, that is surely what we should expect. We have a well-developed intuition about information that has led to the development of a well-established and effective information technology. A scientific theory that produced results counter to our intuitions at an early stage would surely have a hard time being accepted as a theory of information.

On the other hand, intuitions *can* be misleading, and it is always good to be able to confirm one's intuitions on the basis of scientific analysis and logical deduction from "first principles." The point about the development described here is that it provides a unified approach to information that can bear upon work done under many banners: computer science, management science, communications science, cognitive science, psychology, sociology, linguistics, and logic.

## SUMMARY

In previous chapters, we observed the important role played by situations and types, both in the storage, representation, and retrieval of information and in guiding the rational actions of people and other "intelligent" agents. In this chapter, we focused our attention on the abstract links or connections between types that actually result in a particular object or situation encoding or providing information. We called these connections "constraints."

Constraints are what we might call the *regularities* that make intelligent action possible. The constraint that links the *type* of situation in which a traffic light shows red with the *type* of situation in which a motorist stops the car—a constraint enshrined in the traffic laws—gives rise to the *particular* behavior of a *particular* motorist when faced with a *particular* red light.

We used the concept of types to provide a systematic way to view information. Information always takes the form of a statement that some object is of some type. If $a$ is any object and $T$ is a type, we write

$$a : T$$

to indicate that $a$ is of type $T$.

Given a constraint $C$ that links the situation type $S$ with the situation type $R$, $C$ can give rise to the acquisition or transmission of information as follows. If $s$ is a particular situation of type $S$, then the constraint $C$ tells you that there is a situation $r$ of type $R$. The situation $r$ may be equal to $s$, it may be $s$ at some later time, or it may be some entirely separate situation.

Thus, constraints capture within our theory the regularities in the world that give rise to the creation and flow of information.

# 7

# The View
# from Above

-------------------------------------------

## LOOK AT IT THIS WAY

We have developed a way of thinking about information that depends on three fundamental concepts:

◆ situations

◆ types

◆ constraints

In particular, when we want to be precise about information, we view information as always taking the form: some object is of some type, that is, $a : T$, where $a$ is an object and $T$ is a type. We refer to a single item of information of the form $a : T$ as an infon.

To obtain or extract information from some object, you and/or the object need to be in a suitable environment (a situation), and you have to know the constraint(s) that governs the way the object encodes the information. Similar conditions govern the extraction of information by any other "information processor," including a computer.

Suppose that $C$ is a constraint that enables object $a$ to represent the information $u : T$. Then $C$ provides a link between two types:

◆ the type of situation (the situation that will be the context for the representation) in which an object $a$ (the object that represents the

information) has a certain property (the property the object must
have to represent that information);

◆ the type of situation in which another object *u* (the object about
which the information is represented) has type *T*.

For example, imagine the case of a list of names and addresses
stored on a computer disk. How does the disk contain the information about the people listed?

The disk, let's call it *D*, consists of a thin sheet of plastic coated
with a film of magnetic material. The pattern of magnetization in
the magnetic film somehow encodes the information we are interested in. On its own, however, the disk is of no use to us as a source
of information. To obtain the information encoded on the disk, we
need to insert the disk into the appropriate computer.

The computer can discern certain kinds of patterns in the magnetic coating on the disk. In particular, it is designed to provide a
collection of constraints that link certain magnetic patterns on the
disk with certain patterns projected onto the screen or sent to the
printer to be inked onto paper. Given a specific magnetic pattern of
the appropriate type, the computer produces a specific pattern of a
certain type on the screen or on the page. Possible output patterns
might be:

> Alice Bloom     43 Canary Lane, Sidcup
> David Eagles    72 French Close, Sidcup

The computer user is similarly equipped with a collection of constraints that link certain patterns on the screen or the page and certain mental patterns (which we shall call "information"). The user
can use the appropriate constraint to obtain specific information
from the specific pattern on the screen or the page. The user's constraints are those governing the ability to read English when presented in the form used by the computer for its outputs. The information the user obtains will be of the form: Alice Bloom lives at 43
Canary Lane, Sidcup; David Eagles lives at 72 French Close, Sidcup,
and so on.

In our "official" framework for treating information, the information on the disk will be:

> [Alice Bloom] : [type of a person who lives at 43 Canary Lane,
>                 Sidcup]
> [David Eagles] : [type of a person who lives at 72 French Lane,
>                 Sidcup]

and so on. The brackets are just to emphasize the two separate objects on either side of the colon, the individual and the type.

Of course, for types of this kind, different individuals may have the same type (that is, they may live at the same address) and, over his or her lifetime, a single individual may have more than one type (people often change their places of residence).

Notice the way this example works: from the constraint that connects two types to the specific link between one specific situation and another.

## THE SYSTEM APPROACH

With our basic framework now in place, we can start to apply it to different real-world situations. What the theoretical considerations tell us is this: Whenever we are faced with analyzing information flow, we need to shift our focus from a concentration on the words being spoken or the message being sent to a broader view that also includes:

◆ the context situation(s);

◆ the object or situation about which the information is being transmitted;

◆ the constraints (and the associated types) being used to transmit the information.

This is very much a systemwide view. In this respect, what I am saying is neither new nor unique. In recent years, a number of analysts have begun to advocate a systemwide view of business. Perhaps the most well-known proponent of this view in the area of business management is Peter Senge in his book *The Fifth Discipline*. One thing that is different with the approach I am adopting here, of course, is that the systemwide view to information is not being proposed simply as "a good idea"—or even, the more cynical might say, as "the currently fashionable idea." Nor is it based (solely, or even primarily) on a reflective analysis of business practice, as for instance is Senge's book. Rather, it is a consequence of a scientific analysis of information. (A workplace analysis did form part of the basis for the scientific analysis, however.)

In other words, I am not advocating the adoption of a systemwide view of information just because I happen to think it is worth a try, or on the basis of a workplace analysis; rather, adoption of the systemwide view is what the scientific analysis tells us to do!

Moreover, the scientific analysis does more than tell us we should view information at the system level. It shows us how to proceed, namely:

1. Identify what it is that the information is about and what that information tells us about that entity.
2. Identify the critical contextual situations.
3. Identify the constraints that support the encoding and the transmission of the information.

Thinking about information in this way may seem complicated. The reason is that we are so used to tacitly identifying information with a particular representation. Often, when we say "information" we really mean some particular representation. Such an approach can work, and it sometimes does. But it can go expensively or even fatally wrong. (We'll see an example later in the chapter.) By adopting the systemwide view, involving situations, types, and constraints, we can start to develop much more effective, and reliable, procedures to manage information.

How do we set about a systemwide analysis of the kind I am advocating? Given the complexity of many information exchanges—and the ease with which things can go wrong even in relatively simple-looking exchanges—it can often be useful to start by drawing a diagram. By adopting an "overhead view," with different situations depicted as ovals, you can create a diagram that shows the contextual situations and the object or situation being discussed.

One advantage of having (such) a diagram is that it makes clear what are the main channels of communication and what contexts are involved. To illustrate the idea, let's take a look at the most familiar, and indeed the most common, kind of information transfer: the two-person conversation. A conversation can be illustrated by what I call a "conversation diagram" (or more generally, a "context diagram," for reasons that will become clear momentarily).

## THE GEOMETRY OF CONVERSATIONS

In Figure 7-1, the two participants, A and B, are denoted by dots. Their contexts are represented by two ovals, labeled $C_A$ and $C_B$, respectively. These are the participants' contexts for the conversation.

For later use, let me note that the two contexts include a whole range of knowledge, skills, and experience that each participant

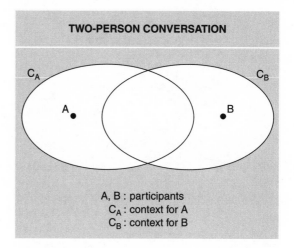

**TWO-PERSON CONVERSATION**

A, B : participants
$C_A$ : context for A
$C_B$ : context for B

**Figure 7-1** Two-person conversation.

brings to the conversation and which contribute to the conversation but which are not themselves part of the conversation. Linguists often refer to these parts of the contexts as the "background" for the conversation.

By drawing the two contexts to overlap, we can use the overlapping portion to denote the knowledge and skills (including the background) that both participants share. For example, in a conversation conducted in English, both participants would be expected to have a command of the English language. That constitutes background knowledge, and will be in the common region where the two contexts overlap.

The nonoverlapping portions of the two contextual regions denote information or skills possessed by one participant but not the other. For most conversations, the aim is for each participant to take information in his or her own context and put it into the overlapping part: They tell each other new things. When A tells B something that B did not already know, then that piece of information, which until then was in $C_A$ but not $C_B$, becomes part of $C_B$ as well as of $C_A$. Thus, we can view a conversation geometrically as a gradual pushing together of the two contexts, so that the overlapping portion becomes larger.

The diagram can be extended to show the object or situation being discussed—what situation theorists usually refer to as the focal object/situation. This is done in Figure 7-2 on the following page.

It would be possible to include in the diagram the relevant types

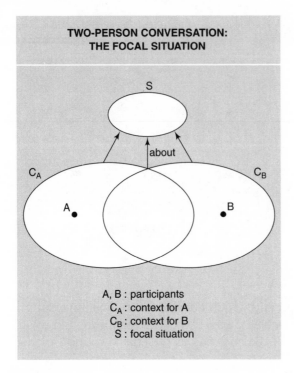

**TWO-PERSON CONVERSATION:
THE FOCAL SITUATION**

A, B : participants
$C_A$ : context for A
$C_B$ : context for B
S : focal situation

**Figure 7-2** Two-person conversation: The focal situation.

and constraints as well, but in most cases this is not done. It would add clutter, with very little gain in return.

Diagrams such as the ones in Figures 7-1 and 7-2 are useful in depicting the participants in a conversation or information exchange. Diagrams that show constraints are more useful when you want to discuss the actual information transmitted.

Before we go any further with our look at conversation, I should observe that viewing conversation as an exchange of information is just one of several ways to analyze the event. Certainly people do use language to convey information, and it can be argued that in a business environment that is often its main purpose. But language may be used for other purposes: to affect the behavior of another, to create empathy, to convey emotion, to establish a position of superiority or equality, and so on. While each of these uses involves an exchange of information, that is not the primary purpose. Although the situation-theoretic approach to conversation adopted in this book takes the exchange of information as the principal focus of the analysis, it is often important not to ignore some of the other aspects of conversations.

One feature of conversations that should not be overlooked is their collaborative nature. A conversation is not simply two people talking in turn. As numerous studies have shown in recent years, conversations have important features that cannot be understood by looking at the behaviors of the two participants on their own. A conversation is an example of what is generally called a "collaborative act" or sometimes a "joint act."

A collaborative (or joint) act is not simply two individual acts performed at the same time. There has to be coordination. For instance, playing a duet on the piano is a joint act. The two players in a duet are indeed performing individual actions—each is playing a piano. But in order for the result to be recognizable as a duet they have to be playing as if they were one. Shaking hands is another example. You can't do it alone. Even if you move your arm and hand in exactly the way you would when shaking hands, it will not look like a handshake, it will not feel like a handshake, and it will not be a handshake (nor even one half of a handshake).

Joint acts are intriguing. They provide examples of phenomena where the whole is greater than the sum of the parts. Two people dancing a tango is another example. As the old saying goes, "It takes two to tango." A well-coordinated tango requires that the two partners move in perfect coordination, each one receptive and responsive to the other's every movement.

One of the main tasks involved in analyzing conversation is to see how it is that the contributions of the two participants fit together to produce a single, joint communicative act. What are the basic steps in the verbal tango we call conversation, and what are the rules of the dance that govern the way those verbal steps are put together?

Of course, none of us ever sets out to learn how to engage in conversation in any conscious way—by learning the rules. We simply do it, starting at a fairly early age. The aim of research into the structure of conversations is not to improve our ability to engage in successful conversations—though much of that research was motivated by the goal of human-computer communication. Nevertheless, an awareness of what is involved in conversation can be very valuable in management. Miscommunication can be costly, sometimes fatal, and it is probably the case that many (maybe even most) failures in communication arise because the participants in a conversation fail to observe the unwritten rules of successful conversation—something they may well not do when engaged in a conversation outside the work environment. Providing facilities and developing procedures to support successful workplace communication could well be one of the most dramatically cost-effective ways for a company to improve its productivity.

I'll finish this chapter by using a context diagram to analyze a dramatic, and tragic, example of miscommunication. It's another airline disaster. My use of such examples does not mean that airlines are unduly afflicted by poor information management. On the contrary, they have a far better record of reliable communication than most other industries. Rather, airlines record all of the crucial exchanges between the aircraft crew and the ground control personnel. In consequence, on the rare occasion when there is a major mishap, we have a publicly available record of everything that was said. Thus, airline disasters provide the information scientist with excellent data on which to test the latest theory.

## THE STORY OF FLIGHT AA 965

In December 1995, American Airlines Flight 965 from Miami to Colombia was on its final approach to Cali airport when it crashed into a nearby mountain range, killing the 159 passengers and crew on board.

When the report of the airline's investigation was made public in August of the following year, it became clear that the problem was not mechanical. Nor was the weather a factor: There was a lot of cloud, but with modern navigational systems that is not a problem. The principal culprit was information—more precisely, the distinction between the little-i information provided by the onboard computer system and the big-I Information on which the crew based their decisions.

Here is how a senior executive of the airline (its "chief pilot") subsequently described the events that led up to the crash. The air traffic controller at Cali instructed the crew to fly toward a nearby beacon called "Rozo," identified on navigational charts by the letter R. The crew entered that letter into the onboard flight management computer, whereupon the screen responded with a list of six navigational beacons. By convention, such a list normally presents the beacons ranked from nearest to farthest from the plane. Since the crew was on the final approach, they did the customary thing and accepted the top entry on the list. It should have been the Rozo beacon, in agreement with the convention used on the printed charts. It was not.

Unknown to the crew, the R at the top of the list actually signified a beacon called "Romeo" in Bogotá, more than 100 miles away and in a direction more than 90 degrees off course. Once the crew had selected the R-beacon on the computer, the autopilot silently and obediently did as instructed, and slowly turned the plane left toward Bogotá. By the time the crew realized something had gone very wrong, it was too late.

Officially, there was no question as to who was at fault. It was the crew's job to know where the plane was headed and what the autopilot was doing. But when alert, well-trained personnel can make such a catastrophic error, we should try to see what circumstances led them to do so. Such mistakes are usually oh-so-understandable and of the "there-but-for-the-grace-of-God-go-I" variety. Once the problem has been identified, ways can be found to avoid a repetition of the disaster.

In the case of Flight AA 965, one obvious and vital factor is the importance of consistent and accurate data. On the charts, the Rozo beacon was labeled "R". But in order to retrieve its listing from the computer, the crew would have had to type the entire word "Rozo". The airline's accident report did not explain the discrepancy, but the chief pilot's report noted "charting and database anomalies that have been discovered."

A more general issue is how to present important information to those who need it. Given the myriad duties of an airline crew as they come in to land, all of which have to be completed in "real time," they have little if any opportunity to double-check every single detail of the masses of information available to them. Aware of the dangers of providing busy crews with too much information, cockpit displays are designed to supply only the really essential information and moreover to do so in as simple and concise a manner as possible. Likewise, it makes sense to arrange things so that the crew do not have to enter a complete word if one or two letters is sufficient.

As an American Airlines' spokesman explained, the screens in the cockpit show only the beacons' code letters and geographic coordinates. Since the corresponding charts generally show those coordinates in print so tiny a busy crew is unlikely to check them—indeed, they may even omit them entirely—the crew will almost certainly rely on the letter or name used to identify the beacon.

Figure 7-3 on the following page indicates the source of the problem in terms of contexts. The crew were operating within two different sets of conventions. The problem was, they were not aware of the fact. They thought they were operating under the normal conventions for abbreviating the names of landing beacons. In other words, the context for their actions was the situation labeled $C_A$ in Figure 7-3. According to the conventions in that context, the letter $R$ denoted the Rozo beacon. Their assumption was that the local computer system used the same convention. If that were the case, then the letter $R$ would denote the Rozo beacon on the computer as well, as indicated by the dotted arrow in Figure 7-3. However, the people who programmed the local computer system had a different context, $C_B$, in which the letter $R$ denoted the Romeo beacon.

Of course with hindsight, it is easy to see how to minimize the likelihood of a repeat of the Flight 965 accident: either make sure that the abbreviations used by the computer system are the same as those on the printed charts, or else arrange for the cockpit display to show the complete beacon name ("Rozo" or "Romeo" in the case in question) along with the coordinates. Drawing a diagram such as Figure 7-3 does not, in itself, solve the problem. What the diagram does do is highlight exactly where the problem lies, namely the context. After all, there was only one crew, only one computer system, and only a single keystroke at the crucial moment, the letter $R$. The two contexts, $C_A$ and $C_B$, were largely in agreement (indicated by their being depicted as overlapping ovals). But one crucial difference

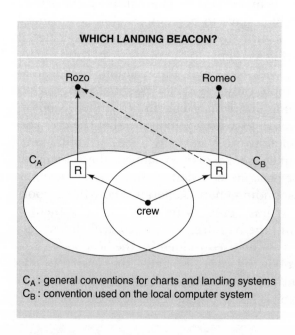

WHICH LANDING BEACON?

$C_A$ : general conventions for charts and landing systems
$C_B$ : convention used on the local computer system

**Figure 7-3** Which landing beacon?

between them was the actual beacon associated with the letter *R*. The convention that made this association should have been in the overlapping region, but it was not.

Leaving aside issues of legal fault and responsibility, and concentrating instead on the flow of information, the cause of the crash can be viewed in terms of the distinction between little-i information and big-I Information. The autopilot did not malfunction. Having been instructed to head for the beacon associated with the letter *R* in its database, it steered the plane toward the Romeo beacon in Bogotá. The little-i information, which was transmitted in context $C_B$, flowed perfectly. (Flight computers don't attach any meaning to the symbols they process, of course. They simply process those symbols as instructed. In this case, selection of the R entry by the crew led the autopilot to send signals to the control mechanisms that had the effect of turning the plane toward Bogotá.)

So too did the big-I Information flow perfectly—in context $C_A$. The problem was that the big-I Information the crew initially attached to the symbol *R* on the cockpit display screen—arguably with every justification—was that it referred to the Rozo beacon just ahead of them, not the Romeo beacon 100 miles off to the left. Through the electronic processing of its little-i information, the autopilot acted as if the letter *R* meant "Romeo" whereas the crew (operating with big-I Information) took the letter *R* to mean "Rozo." As a result, 159 people lost their lives.

### SUMMARY

Information can only be properly understood at the system level. To analyze the way information flows in a system, you have to:

◆ Identify the critical contextual situations;
◆ Identify what it is that the information is about and what that information tells you about that entity;
◆ Identify the constraints that support the encoding and the transmission of the information.

Similarly, to understand the way information is transmitted in a conversation, you have to take account of (1) the contextual situations, (2) the object or situation about which the information is being transmitted (the focal object/situation), and (3) the constraints (and the associated types) being used to transmit the information.

The participants, the contextual situations, and the focal object/situation can be depicted in a diagram in which situations are represented by ovals.

In the case of a two-person conversation, the basic contextual situations for the two participants include a range of relevant background knowledge and skills that each brings to the conversation.

The contextual situations can be represented by ovals on a "conversation diagram." The overlapping portion of the (context) ovals represents the context (including the background) that the two participants share. As the conversation proceeds, the two ovals gradually overlap more, as information is moved from the nonoverlapping parts of the context ovals into the shared, overlapping part.

The conversation diagram captures just one aspect of a conversation. An important feature that the diagrams do not represent is the collaborative nature of conversations. This topic is taken up in the next chapter.

# 8

# Building Walls

-----------------------------------------------------

## THE FALLACY OF THE HOLLYWOOD CONVERSATION

Probably the biggest mistake people make when thinking about conversation is to assume what I call the "He said, she said" model. This model is typified by the highly stylized—and very unrealistic—portrayals of conversation you see in television dramas and in the movies. In those scripted and rehearsed scenarios, first one person speaks, then the other responds, then the first one speaks again, and so on. Each takes his or her turn to speak. There are no false starts, no overlaps, and very often, both participants speak in complete sentences. Other than staying in close physical proximity and possibly facing each other, there is little sense of any *collaboration*—of both participants working to ensure that each individual utterance succeeds.

In terms of information flow, the model portrayed by such scenarios is of neat little packages of information being passed from one person to another like a tennis ball being lobbed back and forth across the net. Person A speaks. This transmits information item 1 to person B. Then person B speaks, transmitting information item 2 to A. Then A speaks, providing B with information item 3. And so on. Some of these items of information may be in the form of commands or questions, of course; they don't all have to be factual. But the information flow portrayed by those Hollywood-style conversations is very much such a tennis ball exchange.

Why is it that such portrayals of conversation don't strike most of us as artificial? Because they are practically the only ones we ever observe. Most of us *participate* frequently in conversations, of course. But, other than on TV or in the movies, we rarely *observe* a conversation—from the outside, as it were. That means we rarely observe a *real* conversation. Most people, when played an audiotape of a real conversation, or when presented with a transcript of a conversation—say, from legal testimony or a congressional hearing—are quite shocked to observe how disjointed it is.

The most obvious thing about a real conversation is that complete, grammatical sentences are rare. Moreover, there are false starts, overlaps, and interruptions. Look a little deeper and you see that, most of the time, we underspecify what we want to convey. We can do so because a conversation is not a sequence of individual utterances but is instead a joint act where both parties contribute. When A is speaking, B is providing constant feedback that A monitors. As a result of the feedback, A is often able to say far less than one might initially assume in order to convey a particular message.

In fact, the most useful way to view conversation—at least, a successful conversation—is as a process of *negotiation:* The two participants negotiate the meanings of the words they utter. Virtually no words of the language are totally unambiguous, and whatever we say can be misunderstood. The aim of a conversation may well be to exchange information, but what actually passes back and forth between the two participants are *representations:* words, phrases, and sentences. In order to achieve a successful exchange of information, the participants have to work together to ensure that both attach the same information to the words that are exchanged (including having the same referents). In terms of information, they negotiate so as to try to ensure that the information conveyed by the speaker's words is the information he or she intends to convey.

Because we all develop the ability to engage in conversation at a very early age, and because we are able to do it so easily, we generally overlook the intricacy of even the most mundane conversations. Somehow, we generally manage to convey our thoughts to others by exchanging words. Analyzing how that is done can help us to avoid those situations where it goes wrong (sometimes with disastrous consequences, as in the case of the Charge of the Light Brigade, discussed in Chapter 5).

To give another example of how easy it is for meanings to go awry, imagine you are an air traffic controller at an airport on a dark, stormy night with poor visibility. (This example is a simplification from an actual event.) You radio the pilot of an incoming plane with the question

*Are you holding?*

The pilot confirms with

*Roger, we're holding.*

Satisfied that the plane is continuing to circle around at its present altitude in a holding pattern, you tell the pilot to await your further instruction and you turn your attention to other planes under your guidance. Unfortunately, the pilot took the exchange to be a confirmation that the plane was holding its already established rate of descent as it approached the airport ready to land. Unless one of you checks the instruments to see the plane's altitude—and in difficult circumstances pilots and ground controllers have been known to overlook this—you have just guided the aircraft toward a crash. All because you were unsuccessful in establishing an agreed-upon meaning of the word "holding": Does it mean holding altitude or holding rate of descent?

As a result of examining large numbers of real conversations, linguists have developed a framework for studying conversation. Instead of regarding a conversation as made up of utterances, they say, you should think of it as a series of joint acts, which they call "contributions."

Contributions are not the same as utterances. An utterance is an act performed by one person. An utterance could be made in the absence of any recipient. A contribution, in the new technical sense, is a joint act that cannot be performed by one person alone.

In the simplest case, a contribution to a conversation is a joint act in which the current speaker (the "contributor") makes an utterance and the listener provides confirmation that the utterance has been adequately understood. In order to proceed, the speaker must obtain positive confirmation from the listener that his or her utterance has been adequately understood. In the absence of such confirmation, the speaker will generally assume a communication breakdown has occurred and will try to correct it. This means that even no action on the part of the listener amounts to a response, which the speaker will note. In a joint act, there is no possibility for either partner to opt out.

The participants in a conversation continually monitor the course of the discourse, looking for confirmation that they may continue, and this is in part what makes conversations collaborative joint acts. In making a contribution, the contributor and his or her partner will work to ensure that they mutually believe that the partner has understood what the contributor meant, to a degree adequate for their current purpose.

Linguists use the term "confirmation devices" to refer to the methods the listener uses to indicate that he or she has correctly, or adequately, understood the utterance. In the most straightforward cases, a contribution to a conversation will involve an utterance plus a confirmation device.

There are a number of confirmation devices that people regularly use in the course of a conversation. The ones listed below are among the most common. Since we all use these devices all the time, none of them will come as any surprise. The reason for spelling them out is that we are trying to come to grips with the key features that make conversation possible. Assume Alice is the speaker, Brian the listener. Thus Brian uses the confirmation device to indicate to Alice that she may continue.

**Continued attention.** Brian shows he is continuing to attend, and thereby signals that he is satisfied with Alice's utterance thus far.

**Acknowledgment.** Brian nods or says "uh-huh," "yeah," "right," or the like.

**Facial expression.** Brian raises an eyebrow to indicate surprise, or shows puzzlement or confusion. In the case of a raised eyebrow, Alice may infer that her utterance has been understood. If Brian shows puzzlement or confusion, Alice will realize she has not made herself fully understood, and will act accordingly, perhaps paraphrasing what she has said or else offering an explanation.

**Initiation of a relevant next contribution.** Brian utters a sentence, or chain of sentences, that makes a relevant and appropriate contribution to the discourse. This includes obvious signals of failure, such as Brian asking for clarification. In such a case, the appropriate response for Alice is not to continue as before, but to provide a repair or clarification of the utterance that Brian has not adequately understood. For instance, if Brian says "Pardon?" or "What was that?" Alice is then obligated to repeat her original utterance, possibly with some rephrasing. Alternatively, if Alice's utterance consisted of her asking a question, then the utterance by Brian of an appropriate answer constitutes confirmation that he has understood Alice's original utterance.

**Demonstration.** Brian demonstrates all or part of what he has understood Alice to mean, perhaps by presenting a paraphrase. Alternatively, an appropriate action by Brian indicates that he has understood. For example, if Alice says "Pass the salt," then Brian's performance of the physical action of passing the salt indicates understanding.

**Completion.** Brian completes Alice's utterance, or rather what he takes that utterance to be. This requires confirmation or rejection by Alice. Confirmation by Alice would normally take the form of an utterance such as "Right," followed by Alice initiating another contribution. Rejection by Alice of Brian's completion would normally be followed by Alice attempting to repeat her first utterance, possibly with some rephrasing.

**Display.** Brian displays verbatim all or part of what Alice has said. For instance, a common way to acknowledge that one has understood the presentation of a telephone number is by reading it back aloud.

## CONVERSATION AS BUILDING A WALL

My favorite analogy for a conversation is of two people building a wall.* Before I can develop this analogy properly, I need to remind you of what linguists call the "background" for a conversation: namely the range of experience, skills, and information that each of the two individuals brings to the conversation, which are not themselves part of the conversation, but without which the conversation could not take place.

In general, much of the background is shared by the two participants. For example, the rules of the English language (or at least enough of those rules) constitute part of the shared background for a conversation between two participants carried out in English. (The shared mastery of the language would not normally be regarded as part of the conversation itself.)

Another aspect of the background that supports any everyday

---

*This analogy has one potential drawback. Two people building a wall may be doing so in order to create a barrier that separates them. This is not the kind of wall building I am thinking of. Rather, the picture I want to invoke is of two people collaborating to build a wall for their common good. They stand on either side of the wall, adding brick after brick in a collaborative effort. The wall they construct is a shared artifact, not a barrier between them.

conversation comes from our commonsense knowledge of the every-day world. Even the simplest of conversations will almost certainly depend on this part of the background.

The background an individual brings to a particular conversation is certainly extensive and in many ways impossible to pin down. On the other hand, it does not consist of everything. For example, suppose you and I have a conversation about football. The shared background for our conversation certainly includes a knowledge of the English language as well as our general commonsense knowledge of the world (such as the fact that a ball that is kicked into the air will come down again). But it does not include a knowledge of Russian or Chinese, or experience in chemical engineering, or an ability to windsurf.

In terms of the wall-building analogy, the physical dexterity the two individuals bring to the task is part of the background, and so too is the supply company that delivered the bricks, sand, and cement. All of these items contribute to the building of the wall, and some of them are essential to the task. But none are part of the actual building work.

The foundations of the wall, however, are more than just background; they really are an integral part of the wall. Likewise the act of laying the foundations is an integral part of building the wall. Laying the foundations corresponds to opening the conversation—establishing what the conversation will be about.

Once the foundations have been laid, construction of the wall then proceeds in a step-by-step fashion, as the two persons add one brick after another in a coordinated and cooperative fashion. The bricks in the first row rest on the foundations. Thereafter, each new brick builds upon those that have been laid previously. The attention of the two people building the wall is focused entirely on the wall and its foundations, not on anything in the background.

In an analogous fashion, a conversation between two individuals may be regarded as a process whereby they cooperate to add information to a common pool. The first part of the conversation is to establish some common starting point. Then, one or both participants adds new information that rests upon that common foundation. Thereafter, each new contribution builds upon the aggregate of information that has been contributed up to that point.

The name linguists give to the common information pool for a conversation—the "wall" to which the two participants contribute—is the "common ground" for the conversation. Not only does each contribution to the conversation add new information to the common ground, the common ground provides a context and a resource for each contribution.

For example, going back to our hypothetical conversation about football, knowledge of English and a layperson's familiarity with gravity and Newtonian mechanics are clearly part of our shared background, as we observed, and the conversation could not take place without them. But they are hardly *part of* the conversation. Thus, they are part of the shared *background*. On the other hand, our specific knowledge of football surely is highly relevant to what is said, and it will thus constitute part of the *common ground*.

One of the constraints a speaker must satisfy in order to successfully initiate a conversation is to ensure that the listener is able to identify—that is, to be aware of—the relevant common ground. An investigation of how two people determine the common ground that supports a conversation is one of the main tasks facing present-day linguists.

For example, if John and Alice are both on the faculty at Saint Mary's College, and if they meet in the college cafeteria, John may open a conversation with Mary with an utterance such as:

*The chapel looks much nicer now that the repainting has been done.*

In this case, John's use of the phrase "the chapel" is perfectly adequate to identify—for Mary—the college chapel, since a basic knowledge of the college architecture and the names of the main buildings is part of their relevant common ground as Saint Mary's faculty.

Now, let's suppose that John and Mary also travel regularly to Stanford University some fifty miles away to participate in a linguistics seminar. Thus, relative to a different common ground, one dependent on their familiarity with and possibly their presence on the Stanford campus, John could use the phrase "the chapel" to refer to the Stanford chapel. Indeed, if the two were standing in the Stanford Main Quad, directly in front of the chapel, John could open a conversation with the same utterance as before, only this time in reference to the Stanford chapel. John's sentence is the same, but it refers to a different chapel. What has changed is the common ground.

One of the important features of common ground is that it involves common knowledge. That is to say, the two participants have joint knowledge of the common ground. As in the case of joint action, joint knowledge involves more than just the two individuals having the same knowledge. Not only do the two participants in a conversation have the knowledge in the common ground, in addition they both know that they both have this knowledge.

The fact that the common ground consists of common knowledge means that, in terms of our conversation diagrams, the common ground is properly contained in the overlapping part of the two background or contextual situations. See Figure 8-1. The overlapping part represents the knowledge and skills the two participants have in common. This will include information that each has but where it is not common knowledge to them that they have that information. Of course, information that has been established during the course of the conversation will be in the common ground—it will be common knowledge.

Clearly, any speaker in a conversation has to design his or her utterances so that the listener can readily identify the relevant common ground—and what is to be added to it as a result of the utterance. Indeed, it is part of the collaborative nature of a conversation

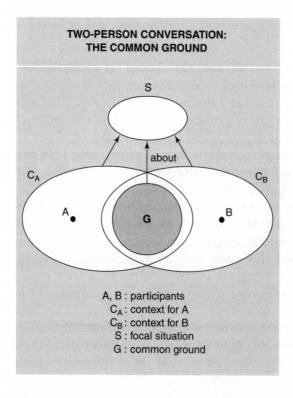

**TWO-PERSON CONVERSATION:
THE COMMON GROUND**

A, B : participants
$C_A$ : context for A
$C_B$ : context for B
S : focal situation
G : common ground

**Figure 8-1** Two-person conversation: The common ground.

that the participants work together to establish and maintain the common ground on which the particular conversation depends. Moreover, both are aware that the conversation creates common knowledge.

The common ground for a conversation can depend on a number of features, including being members of the same community, being in the same physical location, and being referred to in the same utterance.

Examples of community membership are both being mathematicians, both working for the same company, both reading the same newspaper, and both having seen the same movie. If Alice and Brian are in the same physical location, they can make use of that shared location as a physical common ground to refer to any number of items: "this chair," "the clock on the wall," "the temperature," and so forth.

It should be stressed that common ground is local to a particular conversation. John and Alice may have a great deal in common: They are both Americans, they were both brought up and educated in California, they both support the San Francisco 49ers football team, they both own a house in San Francisco, they both have children at the same elementary school, they both read the *San Francisco Chronicle*, and so forth. Each one of these commonalities in their lives may provide part of the common ground for a conversation; more likely, a number of them may contribute to that common ground.

Linguists use the term "audience design" to refer to the way in which speakers construct their utterances to be appropriate for their intended audience (or audiences). In the case of two-person conversations, audience design means that the speaker ensures that his or her utterance can be readily understood by the recipient. In simple terms, what this means is that a speaker should make utterances that the listener can readily understand, choosing the words he or she uses in order to achieve this goal.

For an example of audience design, an utterance of the sentence

*He did a Richard Nixon to the tape of the meeting.*

would be perfectly appropriate for an American listener who was an adult in the 1970s. Any such person would know that this meant erasing part of the recording in a deliberate attempt to hide the truth. However, an utterance of the linguistically similar sentence

*Johnny did an Aunt Alice to the cream cake.*

would not be appropriate for such a general listener. Such utterances are not at all uncommon, but they are generally restricted to conversations within a family, where the listener knows just what it was that Aunt Alice did—say, sitting on a dessert on some occasion in the past.

One final remark: It is often not at all clear whether some particular collection of knowledge or skills is in the common ground or simply part of the shared background. To the participants in the conversation, the issue does not arise, of course. They simply get on with the business of conducting the conversation. The concepts of background and common ground are analysts' inventions. In fact, it is misleading to think of the background and common-ground situations as cleanly delineated regions separated by a clear border. Situations simply provide us with a means to impose some order on a human activity for the purpose of analysis. Since the reality we are trying to analyze is fluid and plastic, our analytic apparatus has to reflect that fact.

## Summary

The naive conception of a conversation as a series of alternating, individual utterances is not what generally happens in practice. For one thing, there are usually false starts, overlaps, and interruptions.

Moreover, a conversation comprises, not a series of individual utterances, but a sequence of cooperative events called contributions. These contributions are mininegotiations that serve to establish the meaning of the words spoken and hence the information they convey.

Typically, a contribution consists of an utterance by one participant followed by a confirmation from the other. The more common confirmation devices are continued attention, acknowledgment, facial expression, initiation of a relevant next contribution, demonstration, completion, and display.

The participants base their contributions on the common ground and design their contributions to add new information to the common ground. Thus, the entire conversation can be regarded as a process of negotiating the identification and the growth of the common ground. This is analogous to two persons setting out to build a wall: They have to agree to work together to build a particular wall, and then they must coordinate their actions to achieve that goal. Each new brick is added to that part of the wall that has already been built.

The participants may (and often do) make use of elements of the common ground in order to communicate their thoughts to the other. Making contributions that are appropriate for the other participant, given what has already transpired in the conversation, is known as "audience design."

# 9

# The Hidden Rules of Everyday Conversation

## PAUL GRICE'S MAXIMS

In 1967, the British philosopher and logician H. P. (Paul) Grice gave a lecture at Harvard University in which he tried to do for conversation what Euclid had done for geometry: namely, write down the axioms from which everything else follows. That is to say, Grice first formulated a list of general principles that govern successful information transmission by conversation; then he used his principles to show how we can often say one thing and mean another.

Perhaps I was being a bit overzealous in likening Grice's conversational principles to Euclid's axioms. Grice was as aware as the next person that human conversations are nowhere near as precise as the propositions and proofs of Euclidean geometry. But conversations are not quite as free and easy as you might first imagine, and Grice's general principles turn out to be both insightful and very powerful.

Since Grice was making *observations* about successful conversations, as opposed to trying to write down rules that participants in a conversation are supposed to follow, he did not call his observations "rules," "laws," or "axioms"; instead, he used the term "maxims." His point was that participants in a conversation follow these maxims *implicitly*—they act *as if* they were following the rules.

Some years after his Harvard lecture, Grice was persuaded to publish the text of his talk as an article, choosing the title "Logic and Conversation." Though well known to linguists and logicians, Grice's article has received remarkably little attention among many of the communities for whom successful conversation can be crucial. The most likely explanation is that most people assume that if there is one thing they can do successfully, it is engage in conversation. Unfortunately, the evidence shows otherwise. Thinking that you have got your message across is common; actually doing so is less frequent.

Grice's work predates situation theory by more than a decade, but it fits in perfectly with the situation-theoretic view of conversation developed in the previous chapter. Certainly, no study of conversation is complete without a look at what Grice had to say.

Grice sets out by observing that a conversation is a cooperative act into which the two participants enter with a purpose. He captured the cooperative nature of conversation by what he called the "cooperative principle":

> *Make your conversational contribution such as is required, at the stage at which it occurs, by the accepted purpose or direction of the talk exchange in which you are engaged.*

In other words, "Be cooperative." In the previous chapter, we captured this norm in our metaphor of conversation as building a wall.

The cooperative principle is fairly general. Grice's next step was to derive more specific principles—his "maxims"—by examining consequences of the cooperative principle under four different headings: quantity, quality, relation, and manner. I shall illustrate these four headings by means of nonlinguistic examples:

**Quantity.** If you are helping a friend to repair his car, your contribution should be neither more nor less than is required; for example, if your friend needs four screws at a particular moment, she expects you to hand her four, not two or six.

**Quality.** If you and a friend are making a cake, your contributions to this joint activity should be genuine and not spurious. If your friend says he needs the sugar, he does not expect you to hand him the salt.

**Relation.** Staying with the cake-making scenario, your contribution at each stage should be appropriate to the immediate needs of the activity; for example, if your friend is mixing the ingredients, he does not expect to be handed a novel to read, even if it is a novel he would, at some other time, desire to read.

**Manner.** Whatever joint activity you are engaged in with a friend, your partner will expect you to make it clear what contribution you are making and to execute your contribution with reasonable dispatch.

In terms of conversation, the category of quantity relates to the amount of information the speaker should provide. In this category, Grice formulated two maxims:

1. Make your contribution as informative as is required.
2. Do not make your contribution more informative than is required.

Under the category of quality, Grice listed three maxims, the second two being refinements of the first:

3. Try to make your contribution one that is true.
4. Do not say what you believe to be false.
5. Do not say that for which you lack adequate evidence.

Under the category relation, Grice gave just one maxim:

6. Be relevant.

Grice observed that it would take a great deal more study to come up with more specific maxims that stipulate what is required to be relevant at any particular stage in a conversation.

Finally, under the category of manner, Grice listed five maxims, a general one followed by four refinements, though he remarked that the list of refinements might be incomplete:

7. Be perspicuous.
8. Avoid obscurity of expression.
9. Avoid ambiguity.

10. Be brief.
11. Be orderly.

As Grice himself observed, his maxims are not laws that have to be followed. They are not like mathematical axioms. If you want to perform an arithmetical calculation in a proper manner, you have to obey the rules of arithmetic. But anyone can engage in a genuine and meaningful conversation and yet fail to observe one or more of the maxims Grice listed. The maxims are more a matter of an obligation of some kind. In Grice's own words: "I would like to be able to think of the standard type of conversational practice not merely as something which all or most do *in fact* follow, but as something which it is *reasonable* for us to follow, which we *should not* abandon" [emphasis as in the original].

## WHEN WHAT WE SAY IS NOT WHAT WE MEAN

One of the more interesting parts of Grice's analysis is his discussion of the uses to which people may put his maxims in the course of an ordinary conversation. In particular, he made successful use of his maxims in analyzing a widespread conversational phenomenon he called "conversational implicature." This step is analogous to using Euclid's axioms to deduce propositions about triangles and circles.

Conversational implicature is when a person says one thing and means something other than the literal meaning. For example, suppose Naomi says to Melissa, "I am cold," after Melissa has just entered the room and left the door wide open. Literally, Naomi has simply informed Melissa of her body temperature. But what she probably means is "Please close the door." Naomi doesn't actually say this; rather it is implicated by her words. Grice used the word "implicate" rather than "imply" for such cases, since Naomi's words certainly do not imply the "close the door" meaning in any logical sense. Assuming Melissa understands Naomi's remark as a request to close the door, she does so because of cultural knowledge, not logic.

We use conversational implicatures all the time. They can be intended by the speaker, or can be made by the listener. For example, suppose Mark meets Naomi and says, "How is the car your brother lent you?" Naomi replies, "Well, it hasn't broken down so far."

Mark's question seems straightforward enough. What about Naomi's reply? Assuming both Mark and Naomi are obeying Grice's cooperative principle—that is to say, they are engaged in a genuine attempt to have a conversation, and not trying to mislead each other—what are we to make of Naomi's words? Presumably Naomi is implying—in a roundabout way—that she does not expect her brother's car to be in good order. She is implicating this unspoken meaning. Most people in Mark's position would probably take Naomi's reply that way. But what is the logic behind this particular use of language? After all, Naomi certainly does not come out and say "My brother's car is likely to be unreliable."

In terms of the maxims, here is Grice's analysis of the Mark and Naomi example. On hearing Naomi's reply, Mark could reason as follows:

1. Naomi's remark appears to violate the maxim "Be perspicuous."
2. On the other hand, I have no reason to suppose she is opting out of the cooperative principle.
3. Given the circumstances, I can regard the irrelevance of Naomi's remark as appropriate if, and only if, I suppose she thinks her brother's car would be likely to break down.
4. Naomi knows I am capable of working out step 3.
5. Thus Naomi is implicating that her brother's car would be likely to break down.

Of course, few if any of us would actually go through such a reasoning process. But that is not the point. In a similar vein, people rarely consult the axioms of logic when putting forward a logical argument, but that does not prevent a logician from analyzing that argument and checking to see if it is valid by seeing if it accords with the rules of logic. The purpose of an analysis such as Grice's is to provide a scientific explanation, based on some initial assumptions. One way to try to understand how Naomi's words come to convey the meaning they do is to imagine Mark being asked the question, "How did you reach that conclusion?" Most people in Mark's position would probably respond with an explanation something like the one just given, though perhaps much shorter and not using Grice's technical terminology.

Though Grice makes no claim that people have any conscious

awareness of his maxims, his discussion of conversational implicature establishes a strong case that the maxims capture part of the abstract structure of conversation. They do after all enable the linguist to provide satisfactory, after-the-event explanations of a variety of conversational gambits.

According to Grice, a participant in a conversation—say, Bill in conversation with Doris—may fail to fulfill a maxim in various ways, including the following:

1. Bill may quietly and unostentatiously violate a maxim. In some cases, Bill will thereby mislead Doris.

2. Bill may opt out from the operation both of the maxim and the cooperative principle, making it plain that he is unwilling to cooperate in the way the maxim requires. For example, he might say, "I cannot say more. My lips are sealed."

3. Bill may be faced with a clash. For example, he might find it impossible to satisfy both the quantity maxim "Be as informative as required" and the quality "Have adequate evidence for what you say."

4. Bill may flout or blatantly fail to fulfill a maxim. Assuming that Bill could satisfy the maxim without violating another maxim, that he is not opting out, and that his failure to satisfy the maxim is so blatant that it is clear he is not trying to mislead, then Doris has to find a way to reconcile what Bill actually says with the assumption that he is observing the cooperative principle.

Case 4 is the one that Grice suggests most typically gives rise to a conversational implicature.

## IMPLICATURES IN ACTION

Let's take a look at some more examples of everyday conversational implicatures.

For some implicatures, no maxim is violated. For example, suppose Roger drives up to a policewoman and says, "I'm almost out of gas," and the policewoman replies, "There's a gas station around the corner." By the maxim "Be relevant," Roger can infer that the gas station is open.

In contrast to the gas station scenario, the next example involves apparent violation of the "Be relevant" maxim in a very clear way in order to produce the intended implicature.

Arthur says, "Bill doesn't seem to have a girlfriend these days." Susan replies, "He has been spending a lot of time in Denver lately." Susan's response will violate the "Be relevant" maxim unless she intends her reply to implicate the fact that Bill has, or at least she suspects that he has, a girlfriend in Denver, and she wants her remark to suggest that this is the reason for his frequent visits there.

For another kind of example, suppose Greg has been telling Melissa of his intention to visit Europe, and has mentioned that he would like to visit her friend Yannis. He asks, "Where does Yannis live?" and Melissa replies, "Somewhere in Greece." Clearly, Greg was asking for the name of the location where Yannis lives, in order to see if it would be possible to visit him. Hence Melissa's reply violates the quantity maxim "Make your contribution as informative as is required." Assuming that Melissa is not violating the cooperative principle, the conclusion Greg can draw is that Melissa violates the quantity maxim because to say more would require that she violate the quality maxim "Do not say that for which you lack adequate evidence." In other words, Greg concludes that Melissa does not know the city or town where Yannis lives. Indeed, assuming Melissa is being as informative as she can, Greg may conclude that Melissa cannot be more specific than she has been.

People sometimes flout maxims in order to achieve by implicature an information exchange they would, for some reason, prefer not to state explicitly. For example, suppose Professor Alice Smith is writing a testimonial for her linguistics student Mark Jones, who is seeking an academic appointment at MIT. She writes a letter in which she praises Jones's well-groomed appearance, his punctuality, his handwriting, and his prowess at tennis, but she does not say anything about his ability as a student of linguistics. Clearly, Professor Smith is flouting the relevancy maxim. The implicature is that Professor Smith has nothing good to say about Jones's ability in linguistics but is reluctant to put her opinion in writing.

Irony is often achieved by a violation of the quality maxim "Do not say what you believe to be false." For example, suppose Jane has been telling Richard how badly her friend Sally had let her down, and Richard comments, "Well, Sally certainly is a great friend." The implicature is that Sally is a very poor friend.

Metaphor is another linguistic affect that may be achieved by flouting the same quality maxim. For example, if Tom says to his wife, "You are the cream in my coffee," the implicature is that Tom thinks his wife is the completion to his life.

Violation of the quality maxim "Do not say what you believe to be false" may also be used to achieve the effect of understatement. An example of this is where Barbara and George have had an enormous fight, in which Barbara ended up flinging crockery all over the kitchen, and the next morning Barbara approaches George and says, "I was a bit annoyed last night." The implicature is that Barbara was, as George knows full well, thundering mad. In this case, George probably takes her words as an acknowledgment of, or even an apology for, her behavior.

So far, none of the examples have involved the maxims of manner. Here are three that do.

Parents of young children sometimes flout the manner maxim "Avoid obscurity of expression" in order to communicate with each other in a manner that their children cannot comprehend, saying things like "Did you pick up the you-know-what on your way home?"

Politicians sometimes try to violate the "Avoid ambiguity" manner maxim in order to mislead their audience. In a notorious case in Britain some years ago, a prime minister promised not to take a certain action without "the full-hearted consent of the British people." Most people took this to be a promise of a referendum, but in the end the issue was decided by a single vote in Parliament. In the furor that followed, the prime minister pointed out that in a parliamentary democracy this did amount to "full-hearted consent."

Neither of the above two examples results in an implicature. However, suppose John says to Sally, "Mary produced a series of sounds on the piano that sounded like 'Home on the Range'." This violates the manner maxim "Be brief," and the implicature is clearly that Mary's piano playing was not very good.

Here is one final example of implicature. In this one, the "Be relevant" maxim is violated. The scene is a cocktail party at a company headquarters. Not knowing that the company CEO is within earshot, Theresa says, "The CEO really is a pompous ass." Max replies, in a purposeful tone, "Where did you say you are going for your vacation this year?" Provided Theresa has her wits about her, she will realize that Max must be violating the relevance maxim for some reason, in this case to cover the fact that she has just made a terrible professional gaffe. With a bit of luck, and Max's help, she might just manage to escape with her job intact.

## SUMMARY

Grice formulated a set of maxims that describe the way the participants in a conversation should construct their contributions so as to be successful in getting their point across. He based his maxims on a single principle, the cooperative principle. The individual maxims arise by examining consequences of the cooperative principle under four different headings: quantity, quality, relation, and manner.

Grice's maxims can be used to explicate the logic whereby participants in a conversation use words to convey information. One particular device that Grice investigated in this way is conversational implicature, whereby a speaker says one thing and means another.

# 10

# The Art of Successful Conversation

## HOW TO GET YOUR POINT ACROSS

You're faced with the big moment: You have fifteen minutes to make your case—to get that crucial message across to your boss, your potential client, your lover, or whomever. How do you maximize your chances of success?

The bird's-eye view of a conversation depicted in Figure 8-1 (page 88) provides our starting point. First a caveat: Such conversation diagrams depict the contextual situations of the participants as neatly drawn ovals having a clear region of overlap. In reality, the contexts have fuzzy boundaries and an equally fuzzy region in common. Some skills or items of information are definitely in a particular contextual situation and others are definitely not; but there are many skills or items of information for which it is not at all clear whether they are in or out.

For instance, for two people conversing in English about car mechanics, a basic working knowledge of English will be in the shared background, and a knowledge of car mechanics will probably be in at least one person's context. It may be hard or impossible to specify just what knowledge of car mechanics is in the common ground, though any facts established during the conversation surely will end up in the common ground.

On the other hand, a knowledge of Russian will surely not be in either context. It may be that one or both of the participants has such a knowledge, but that knowledge will not be at all pertinent to the conversation at hand.

Thus, it is important to think of the ovals that depict the various situations as having fuzzy boundaries that are merely approximated by the neat lines drawn in a conversation diagram.

Just what kinds of entity you regard as being in the contexts and the common ground situations depends on the kind of analysis you are pursuing. Situation theory highlights the importance of context in general, but the details depend on the specific reason that drives your analysis. If you are a computer engineer designing the interface for an automatic teller machine—and therefore one of the participants in the "conversation" is a computer—you can probably get away with relatively simple contextual situations. (But beware: Even in such seemingly simple scenarios, things can easily go wrong.) On the other hand, if you are planning an important meeting with the boss, then you had better treat the contexts in a more complex fashion. Here's why.

In a conversation between two individuals, where person A is trying to persuade person B of something, A would be well advised to take into account as much of the context situation of B as possible, including B's habits, preferences, motivations, and lifestyle. A's aim is to add something to B's context situation. That generally means A has to work within B's cognitive territory. In so doing, A needs to be careful not to give the impression of trying to *invade* that territory. People are usually very protective of their physical territory, and the same is true of their cognitive territory.

B's actions are governed by the motivations and behavioral patterns in his or her context situation, so for A to be able to persuade B to perform some action, A should seek a way to utilize or modify constraints in B's context—that is, to work *through* B. This undertaking can require that A take time early in the conversation (or before it takes place, when A carries out the advance preparation for the meeting) to identify some of B's guiding constraints—to find out what motivates B and makes B tick.

In terms of the execution of the conversation, the problem facing A is that any information that is successfully communicated to B during the conversation has to be built upon information in B's contextual situation. If A does not already know that necessary information in B's context, then A will have to elicit that information itself during the conversation.

Moreover, the conversation itself has to be built on the common ground. Part of the common ground will be clear to both A and B at the outset; the identification of the other parts has to be established as the conversation unfolds. Feedback—using the various confirmation devices—is particularly important here.

Most of what is said above is done instinctively—at least by people who find they are naturally good at persuasion. But by being aware of the importance of the contextual situations and adopting the view of conversations depicted by the conversation diagram—in particular the territorial aspect and the "wall-building" negotiation of the common ground—everyone can improve their performance.

Of course, all of these considerations are very theoretical. In the next section, we'll see just how it can work in real life.

## BUYING THE FARM

One of the most successful negotiators I have ever heard of was a real estate agent. On one occasion, he was approached by a client who wanted to purchase a particular ranch in New England. The existing owners were an elderly couple, and the ranch had been in their family for several generations. Though the ranch had barely turned a profit for many years, the couple had turned down several attractive offers, each time saying that they would never dream of leaving the family home. The potential purchaser was not interested in the ranch as a business. He had fallen in love with the idyllic location, and wanted to purchase the property to demolish the house and build a new luxury home. He planned to sell off most of the land, keeping just enough to exercise his horses.

The would-be purchaser had approached the couple himself and made a generous offer, but as with all others before him, the couple had said no, they had to intention of selling. Having heard of the real estate agent's reputation, the purchaser approached him in a last attempt to try to secure the property. After a single thirty-minute conversation with the couple, the agent had secured a deal—for less than the purchaser had originally offered!

How did the real estate agent do it? His first step was to do some local research. Neighbors confirmed that the couple were indeed finding it hard to survive financially, though they were also quick to report that there was little chance they would sell. They had no children, and, as far as anyone knew, their only relatives were the wife's sister and her family, who lived in a village several miles away. A bit more research revealed that there was a nice house for sale in that village that would more than meet the elderly couple's needs. At this point the real estate agent had enough knowledge of the couple's background (that is, the background situation for the forthcoming conversation) to proceed.

Part of the background the agent knew about was that the couple habitually said no to any requests to sell. So there was little point in approaching them with an offer up front. In the terms of our scientific framework, the agent knew some of the constraints that guided the couple's behavior—he had discovered both pertinent facts and pertinent constraints in the couple's contextual situation (for the conversation he was planning). In particular, he knew he could not assume a willingness to negotiate a sale as an opening common ground.

The agent decided to engineer a "chance" meeting, with the property itself as the common ground. He drove by at frequent intervals until an occasion when the husband was leaning against a fence looking out over the property—looking at the same stunning view that had attracted the purchaser.

"Nice view," the agent commented, as he stepped out of his car and strolled over to where the man was standing.

"Sure is," came the reply.

"Yours?"

"Uh-huh," the man nodded.

"Lived here long?"

"All my life."

The agent allowed a short pause, before continuing, "Must be hard to make a profit on a place like this these days."

The man agreed, and the two discussed some of the problems facing a small rancher. The rancher did most of the talking; the agent just provided responses to keep the conversation going. He wanted to ensure that the common ground of the conversation contained information about the problems facing the elderly couple as they struggled to keep the ranch going—he wanted to ensure that the conversational wall they were building had the right bricks at the bottom.

"Friends of mine were in a similar position," the agent continued. "In the end, they decided to sell. They got enough to buy a nice little place near their family and have enough left over to keep them comfortably the rest of their days. Never looked back."

The rancher said nothing, but the agent continued to describe the situation with his friends—in fact, the people the agent was referring to were not actually friends but former clients. However, he wanted to keep the tone of the conversation on a human (as opposed to a business) level. He knew he had to stay away from the rancher's behavioral constraints pertaining to selling the property; instead, he wanted to move into the common-ground constraints relating to financial security, ease of life, and a desire to be near to one's family. By monitoring the rancher's reactions, the agent could see when his story was starting to have an effect.

A crucial moment came when the rancher mentioned that there was a property for sale in the village where his wife's sister lived that would be adequate for their needs. Later in the conversation, the agent "mentioned" that he knew of a person who might just be interested in purchasing the ranch—if, by chance, they ever thought of selling. Within a few more minutes, the deal had effectively been struck.

The agent never made an offer to buy; he left it to the rancher to make an offer to sell. Admittedly, the agent used a little subterfuge to set up the deal. He hid the fact that he was an agent, and he described a former client as a friend. Some real estate agents might say that was not ethical. The agent himself sees it differently. "Everyone got what they wanted," he observes. The purchaser got the property he wanted. The agent got his commission (from the purchaser). And the couple got what they wanted (as opposed to what they had kept saying they wanted), namely security.

As the agent had correctly reasoned, the couple had seen any request that they sell their home as a threat to their security, and as a result had always said no. But the agent did not present the situation in that way. He showed them a way to achieve *greater* security— to live in a smaller house that was easier to maintain, to have the financial security of money in the bank, to be free of the worries of running the ranch, and to live near to a close relative. With that scenario in mind, selling the house was merely a step along the way. They were no longer *selling* their home; they were *buying* another home. In terms of the conversation diagram, the agent actually changed the focal situation from the "obvious" one of the rancher selling his existing home—which others had tried and failed—to one of the rancher buying a new home!

The agent almost certainly did not plan this conversation using conversation diagrams and situation theory. He was simply an experienced and instinctively good negotiator, who understood human nature. But we can use situation theory to examine what he did and to help ourselves to plan equally crucial conversations. In brief, what the agent had to do was find a way to "enter" the rancher's cognitive territory and, working in that territory, bring into the common ground of the conversation the important facts and constraints that would lead to the rancher deciding to sell.

## CONFUSION AT THE AUTOMATIC TELLER MACHINE

The agent described in the previous story was a naturally gifted negotiator. In particular, not only did he know how to plan a crucial conversation, he was very good at reading all the feedback signs as the conversation proceeded. Few of us may ever be as good at negotiating as that agent. The hardest part—the part that you can't learn from books—is knowing how to read the other person's responses and react accordingly. Though you can try to anticipate in advance the various ways the conversation might go and prepare a range of possible counterresponses, it is usually very difficult to anticipate all eventualities. I came across a striking example of that some years ago when my wife and I moved to California and started to use one of the West Coast banks.

One day, my wife went to the automatic teller machine (ATM) to deposit a check and draw out some cash. After she had inserted her card and entered her password, the machine asked her what service she required. She chose "Make a deposit". The machine responded with the question "Do you want cash back from your deposit?" My wife answered "Yes". The machine then presented my wife with the instruction "Enter amount". My wife took this to mean the amount of cash she wanted back, and entered "$100.00". The machine responded with "Enter amount you want back." At that stage there was nothing my wife could do except cancel the entire transaction and start again.

What is interesting about this story is that I had been using the same ATM for some time to make deposits and draw out cash without any problem. As a mathematician and computer scientist, I automatically viewed the preplanned, artificially staged "conversation" with the machine as a machine transaction, and as a result I had always taken the machine's instruction "Enter amount" to refer to the first decision point in the process, namely my choice to make a deposit. In technical terms, I assumed the machine operated using a simple model of what computer scientists call a queue—first in, first out. My wife, on the other hand, viewed the transaction as a highly constrained human interaction, where reference would generally be to the thing mentioned last. Since the instruction "Enter amount" came immediately after reference to wanting cash back, she took the instruction to refer to the amount of cash she wanted back. The model she was assuming was that of a computer scientist's stack—last in, first out.

What is especially instructive about this story is that the people designing the ATM, who were trying to anticipate the "conversation" that would take place between the customer and the ATM, did not foresee the confusion that could arise with that instruction "Enter amount". And yet, it is not as if automatic teller machines were designed hurriedly. Because of the importance for ATMs to be extremely reliable, probably more time has been devoted to their development and design than any other device of comparable engineering complexity. In terms of the design of interactive machines, the ATM is state-of-the-art. And yet "simple" errors can occur, as the above example demonstrates.

No matter how obvious such "errors" may seem after they have been discovered, anticipating them all in advance is nigh on impossible. We can't expect situation theory—or any other scientific theory—to eliminate such errors. They will always arise. Rather, the strength of situation theory is that it provides a way to view conversations—either human-human or human-machine—that forces you to be aware of *where* things can go wrong, even if you can't be sure exactly *what* can go wrong.

The engineers designing the ATM probably thought of the customer as a single entity. If, instead, they had drawn a conversation diagram and asked themselves what were the important features of the customer's background and contextual situation, and what were the possibilities for the common ground when a customer used the machine, they might have realized that different customers could interpret the instruction "Enter amount" in different ways. The conversation diagram forces the designer to consider the context.

---

SUMMARY
In planning an important conversation, it can be useful to draw a conversation diagram and ask questions about the various components of that diagram:

- What are the important facts in the two contextual situations?
- What are the important constraints in the two contextual situations?
- What is the right focal situation to start out with?
- What focal situation do you want to end up with?
- How should you engineer the common ground to steer the conversation in the direction you want it to go?

If you are trying to persuade someone to do your bidding or adopt your way of thinking, you should view that person's contextual situation as his or her own cognitive territory. Getting something into the common ground then requires either moving it from the other person's context or else bringing it in from your own context. Either way, you will need the other person's cooperation to get it into the common ground and his or her acquiescence for the eventual acceptance of what you propose.

The saga of the automatic teller machine shows how difficult it can be to anticipate all eventualities in a conversation. Situation theory and conversation diagrams can help you prepare the ground, but you will still need to keep your wits about you and be prepared to think on your feet.

# 11

# Three's a Crowd (Maybe)

## GETTING MINDS TO MEET

In the previous chapters, we have examined the role played by context in conversation and seen how different kinds of contextual features can influence people's behavior. We introduced a simple diagram to depict the role played by context in a human exchange—a conversation diagram (or, more generally, a context diagram). In this chapter, we'll use that device to examine the likelihood of miscommunication when two or more people meet to exchange information.

"Two's company, three's a crowd," goes the old saying. To anyone who has taken a good, hard look at the structure of two-party conversations, nothing could be more true. Despite the intricacy of the sequence of contributions that make up a conversation, the conversation diagram, shown in Figure 8-1 (page 88), provides a clear and simple representation of the important contextual situations that play a role in a two-person conversation.

When A and B converse, their conversation is built upon the common ground, G, for the conversation. The common ground is that body of information and skills relevant to the conversation that is common knowledge to both participants. Being common knowledge means that they both not only share that knowledge and those skills but also are aware that they do so. Moreover, they are both aware that they are both so aware, and so forth, ad infinitum. (This implies that a genuine conversation can only be carried out by two or more conscious entities capable of self-reflection. An interaction between a person and a computer or between two computers will thus not constitute a genuine conversation, and it can be misleading and even dangerous to regard any kind of computer interaction as a conversation.)

During the course of the conversation, each participant may refer back to information in G, and each may utilize information in G in order to make specific references to people, places, events, and the like.

Moreover, the identification of the focal situation, S, is essentially part of G. I skipped over this point when I discussed the conversation diagram in Chapter 7. I simply assumed that A and B know what situation they are discussing, and I drew all arrows pointing to the same situation S. But, as was made clear in dramatic fashion by the Charge of the Light Brigade and the case of American Airlines Flight 965, discussed in Chapters 5 and 7, this is not always the case.

Identifying and maintaining the common ground is the most important single factor that determines whether a conversation is successful or not. The various confirmation devices listed on pages 84–85 are all intended to try to ensure that each new contribution made by a participant is based on information in the common ground.

The common ground is cumulative. If the confirmation devices work correctly, the information each participant contributes goes into the common ground. Later parts of the conversation can build upon what is established earlier.

Thus, if the conversation is an important one, the participants should do everything they can to ensure that neither of them strays beyond the common ground.

## AND THEN THERE WERE THREE

According to some recent studies,* when it comes to certain important tasks (including collaborative problem solving), by far the most successful of all collaborative teams are the *dyads*—the teams of two people. (Of course, even in the case of collaborative problem solving, the dyad will not be optimal if the problem requires several

*Including Panko & Kinney (1992) and Schwartz (1995).

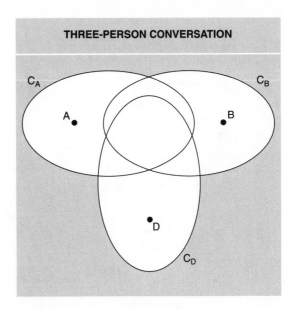

**THREE-PERSON CONVERSATION**

**Figure 11-1** Three-person conversation.

distinct kinds of knowledge and expertise that can only be obtained by assembling a larger team.)

Moreover, the difference between dyads and all other teams is far greater than the difference between, say, three-person teams and teams of four or more persons. There are probably several factors at play here, but almost certainly communication is one of them. In particular, going from a two-person conversation to a conversation involving three or more people is so significant it is probably misleading to continue to use the same word "conversation."

The relative complexity of even a three-person conversation, compared to a two-person conversation, is clear as soon as you try to draw the conversation diagram. Figure 11-1 is the first stage of such a diagram. It shows the three participants, A, B, and D, together with their respective background (or context) situations, $C_A$, $C_B$, and $C_D$.

Though there is a background region shared by all three participants, there are also regions shared by A and B but not D, by A and D but not B, and by B and D but not A. Because the three participants start out with different backgrounds, having different overlaps, there will be three pairwise common-ground situations, one for A and B, one for A and D, and the third for B and D. Figure 11-2 on the following page shows in addition the three two-party common-ground situations, labeled respectively $G_{AB}$, $G_{AD}$, $G_{BD}$.

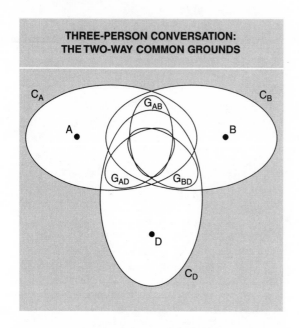

**Figure 11-2** Three-person conversation: The two-way common grounds.

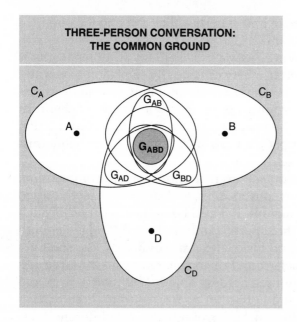

**Figure 11-3** Three-person conversation: The common ground.

Figure 11-3 shows the three-way common ground $G_{ABD}$ for the conversation. This is the information and skill pool that all three participants not only share but also are jointly aware of so sharing.

For the three-party conversation to be a genuine analog of the two-person case, all contributions should be built upon, and contribute to, the three-way common ground $G_{ABD}$.

Already the diagram is too complex to understand easily, and yet I have left off the focal situation. In many cases, the focal situation is sufficiently clear from the start. But just as two people can sometimes get it wrong and unknowingly talk about two different situations, so too it is possible for three people to get it wrong: There may be two focal situations involved, one shared by two participants and the third by the single remaining person, or there may even be three separate focal situations.

And remember, even in the two-person case, a conversation diagram is already a simplification of what actually takes place!

Now, the point of going through the development of the three-person conversation diagram is not to show that conversations with three or more people cannot take place. They obviously do, all the time. Nor is the intention to prove that a conversation involving three or more persons cannot be successful. People have been meeting and discussing issues in groups, sometimes quite large ones, for thousands of years,.

Rather, I want to show just how much more likely it is for a misunderstanding to arise when there are more than two people involved. Not only are there more possible common grounds, resulting from the different combinations of participants into pairs, but the majority of the confirmation devices that guard against misunderstanding in a two-person conversation are not well suited to use in a group of three or more.

You can see how difficult it is to achieve common understanding in a conversation involving three or more people by the fact that, when a conversation has ended, the participants frequently—not just occasionally, but *frequently*—disagree as to exactly what was decided. The larger the group, the more common is such lack of agreement, but it happens whenever three or more people meet.

Sometimes such after-the-event disagreement seems to be willful. Most of us suspect that this is the case when the parties in a labor dispute or a political meeting subsequently give different accounts of what took place and what was agreed upon. But it need not be deliberate. The sheer complexity of multiparty conversations (even two-party conversations) is already enough to lead to major miscommunications. If the parties involved come from different cultures (such as management and labor) or different political ideologies (conservative and liberal or democratic and totalitarian), or if they speak different languages, then the likelihood of misunderstanding is even greater.

Since the number of possible pairs in a group of $N$ people is $N(N - 1)/2$, this formula gives, in crude numerical terms, the number of ways that the common ground can go wrong (that is, can cease to be common to all participants). For $N = 3$, the formula gives the value 3, when there are 4 persons, it is 6, for 5 persons 10, and for 6 persons you get 15 ways things can go wrong.

The possibilities of error are reduced if one person in the meeting is declared the chair and charged with reporting what has been agreed. In order for the chair's report to represent what was genuinely agreed, the chair need only achieve common understanding with each participant. Thus, if there are $N$ people present, including the chair, there are just $N - 1$ ways that things can go wrong. In this case, success depends ultimately on the outcome of $N - 1$ two-party exchanges, one each between the chair and each other participant.

However, now we are getting far too theoretical. As we all know, even when there is a chair charged with making the final report, it is hardly ever the case that the chair talks one to one with each participant to ensure agreement. To do so would take far too much of the chairperson's time. Instead, the wise chair pauses at various stages in the conversation to summarize what she thinks has been agreed and to ask for corrections. Then, when the meeting is over, she provides all participants with a written summary of what she thinks was agreed and asks for individual corrections. Of course, all such corrections should themselves be circulated to all participants in order to ensure agreement on them as well, so this process might have to go through two or three iterations before agreement is reached.

## BUT REMEMBER, TWO HEADS CAN BE BETTER THAN ONE

It turns out there is some business folklore (but no hard evidence, as far as I know) that says three can be a better group size than two when it comes to minimizing the possibility of miscommunication. The reasoning goes like this: Any speaker in a three-person conversation has to work much harder to be understood. After all, what is said needs to be understood by not one, but two listeners, and he or she must confirm with both listeners that they have understood. Even if the bulk of the exchange is between just two of the three participants, then the additional constraint of ensuring that the third person—the "onlooker"—is able to follow the conversation will reduce the possibility of miscommunication. To appropriate the old saying, maybe two listening heads are better than one.

Again, I am not aware of any concrete evidence in favor of this belief, but both the belief and the explanation I have given seem reasonable. Of course, you might be tempted to try to extend the argument, and say that if 3 is a better number than 2, then 4 should be better still. But remember, there is a tension between this caveat argument and our original mathematical argument in terms of maintaining the common ground and the focal situation. There may indeed be circumstances when a three-person conversation produces more accurate information flow than a two-person conversation, but already with four people the complexity of the conversation diagram will surely outweigh the psychological factor of audience design. (Remember, when $N = 3$, the number of possible miscommunications is 3, but when $N = 4$, the number of ways is 6, twice as many.) Again, the common wisdom supports this conclusion—I have never heard anyone claim that four-person meetings resulted in less miscommunication than meetings involving two or three people—but in this case also we await hard evidence.

In any event, in the face of considerable informed opinion, it seems wise to temper our original claim about the informational supremacy of two-person conversations. Instead, we should assert that conversations that involve at most three people are far less likely to lead to miscommunication than conversations with four or more participants.

In the following chapter, I will give another measure by which two- and three-person conversations are better than meetings involving larger groups, namely, the likelihood that the group will exchange and discuss new information rather than simply go over facts that were already familiar to all.

## STACKING THE ODDS

Given the potential difficulties in achieving an acceptable transmission of information (as opposed to the representations thereof) in a conversation involving three or more people, what, if anything, can we do to improve the likelihood of success? (Presumably anything that can help conversations with three or more participants will also help in the case of two-person conversations as well.)

Let's concentrate on what is generally the most significant feature of a conversation in terms of situation theory: the common ground. This is particularly important due to the cumulative nature of the common ground. (Recall the "building a wall" analogy from Chapter 8.) Once participants have lost track of the common ground, they lose the entire remainder of the conversation—they can neither fully contribute nor will they be likely to understand correctly later contributions by others.

Of course, we'll assume that all participants are fluent in the language of the conversation, that they are all reasonably intelligent, that they all have a reasonable level of conversational skill, and that they are all involved in the same business—either they all work for the same company, or they work for two companies engaged in some kind of negotiation, or something else along those lines. By making these assumptions we can ignore many linguistic and social/cultural issues.

In theory it can help to minimize the effects of differing backgrounds and cultures if the meeting starts by having everyone give the group a brief introduction to themselves. However, in my experience, if more than three or four people are involved, this has very little benefit other than perhaps to put everyone at ease. People tend not to listen to introductions of people they already know, and they soon forget or confuse the self-descriptions provided by strangers.

One common and potentially effective strategy is to make regular notes on a whiteboard or a display pad as the meeting proceeds. To do this systematically and to greatest effect, the person leading the meeting should begin by writing up the purpose(s) of the meeting and the assumptions being made, as well as perhaps the way he or she wants the meeting to proceed. Each time a major new issue is brought up or agreement reached on some point, the leader writes that up on the board.

An analyst would say that the whiteboard is an example of a *common artifact*. Common artifacts provide information in such a way that it readily becomes common knowledge to everybody having simultaneous access to it. The whiteboard provides common knowledge because it is a public display, and thus everyone in the room can see one another looking at the board.

The use of a whiteboard is quite different from everyone in the room taking their own notes. You and I may see each other taking notes, but as the meeting proceeds, I cannot be sure what notes you have taken and you won't know what I have written. So, even though we may have taken the same notes, the information in our notes is not automatically common knowledge—that is, writing information in our individual notebooks does not make it common knowledge, as happens when that information is written on a whiteboard. This is the key distinction between a common artifact and a private source of information.

The annotations made on the whiteboard do not need to be complete descriptions. The important thing is that they serve to identify the issues discussed and the agreements reached. Bulleted lists consisting largely of keywords are often all the information that is needed. Simple line drawings, flowcharts, and sketches can also be highly effective. As with the postmeeting written summary, the aim is not to produce complete minutes or a binding legal document but rather to serve as a guide to maintain the common ground.

## SUMMARY

The conversation diagram undergoes a dramatic increase in complexity when you go from a two-person conversation to a conversation or meeting involving three or more people.

This fact probably explains, at least in part (and probably a large part at that), why two-person teams (dyads) are much more effective than larger teams at certain important tasks (including collaborative problem solving, provided the problem does not require a range of specialized knowledge or skills).

There is some unscientific evidence that meetings of three people can sometimes be better than dyadic conversations.

With four or more people in a meeting, identifying and maintaining a single common ground becomes increasingly more difficult.

In meetings involving three or more participants, the use of a common artifact such as a whiteboard can help maintain the common ground. (In fact, even two-person meetings can benefit from the use of a common information storage device.)

# 12

# Going Round in Circles

- - - - - - - - - - - - - - - - - - - - - - - - - - - - - - - - - - - - -

## SO WHAT'S NEW?

Managing the common ground is key to the success of a meeting. However, the larger the group, the more difficult it is to keep everyone on the same common ground. In the absence of common artifacts such as whiteboards, miscommunication is highly likely in a meeting of more than three people. A further argument in favor of two- or three-person meetings comes from the observation that the larger the group, the greater the proportion of time that will be spent discussing information already known to most or all of the participants.

Of course, there are occasions when the recapitulation of known information is exactly what is required. Often, however, the purpose of a meeting is to exchange new information. People want to learn something, to consider new options, or simply to hear what's new. If that is the case, then the tendency of larger groups to focus on information that is already familiar to the majority argues against larger groups and in favor of groups of two or three at the most.

In fact, even for two-person meetings, most of the time will likely be spent discussing information that is already known to both participants, but the time spent on familiar matters will be less than for meetings of three or more persons.

In addition to a number of empirical studies, this phenomenon—the focusing largely on information already familiar to most or all participants—has also been measured in laboratory situations in which groups have been assembled and briefed to discuss a given issue. Stasser (1992) provides a good overview of some of this work. He also considers the points we examine next.

## WHY DO WE TALK ABOUT WHAT WE KNOW?

There are several factors that contribute to the tendency for a group to concentrate on information already known to most or all. One is the psychological factor: Many people simply find it difficult to introduce any item into a group. As a result, there is a tendency to keep talking about what is already in the conversation.

A second factor is group dynamics. Even when a person introduces a new item, if the item is known only to that person, then he or she will have to put considerable effort into getting the group to discuss it. The more people there are who already have the information, the greater the chance that its introduction will lead to a discussion.

On some occasions, participants in a meeting may see their role as advocating a decision they have already reached independently (and thus based on the information they had prior to the meeting). If such is the case, then it can be extremely difficult for anyone to persuade the group to discuss any new information.

A third factor acting against the introduction of new information known only to one or two participants is purely theoretical: It

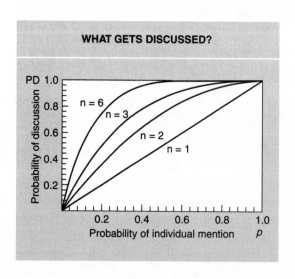

**Figure 12-1** What gets discussed?

is *statistically* unlikely for any item of information known only to a small proportion of group members to be discussed. The reasoning is as follows.

Suppose that $n$ people at a meeting know a particular piece of information $I$ and that the probability that each of those people will introduce $I$ is $p$. What is the likelihood that the group will discuss $I$?

Figure 12-1 shows several graphs relating the probability that item $I$ is discussed (PD) to $p$, for different values of $n$, including the case $n = 1$, where only one person knows $I$.* From the graphs, it is clear that the likelihood that $I$ will be discussed increases with $n$. This effect is particularly marked (the graph rises steeply) for values of $p$ less than 0.4 (that is, 40 percent).

Now, in practice the probability of a specific participant A introducing a particular new item of information $I$ may well be less than 40 percent. For one thing, A will probably have a whole range of new information that he or she could bring up, and $I$ will have to "compete" against all the other items in that range. Moreover, human memory is never perfect, and the many distractions that occur during a meeting can lead to a participant overlooking an item he or she previously intended to bring up. Still another factor is that in many meetings attention is focused on reaching a consensus, and this leads participants to concentrate on understanding and assimilating information already introduced by others. It can also happen that a participant may intentionally withhold information from the others in the group to gain some strategic advantage later. All told then, the probability $p$ of A bringing up the particular item $I$ will most likely be far less than 40 percent.

Indeed, a value of 20 percent or 30 percent is not at all unreasonable. For such a value, the huge disparity between the values of PD for increasing values of $n$ implies that the group is far more likely to spend most of its time discussing items known to most or all of

---

*Here, is how the graphs are obtained. As stated in the text, $p$ denotes the probability that any one person in a meeting who has a particular item of information $I$ will introduce $I$ at some time during the meeting. (For simplicity, we assume that the probability is the same for each of the people that know $I$.) We suppose $I$ is known to $n$ people in the meeting. Then, for each person A who knows $I$, the probability that A does not introduce $I$ is $1 - p$. The probability that none of those $n$ people introduce $I$, and hence that $I$ does not get introduced, is thus $(1 - p)^n$. Thus, the probability that I does get discussed is

$$PD = 1 - (1 - p)^n$$

These are the functions that are graphed for different values of $n$.

them than items known only to one or two. For instance, in a three-person group where each person introduces a third of the information he or she brings to the meeting, the probability that a particular item known only to one person will be introduced is 30 percent, whereas the probability that a particular item already known to all three will be introduced is 66 percent, over twice as much.

The bigger the group, the greater will be the disparity, and the less likely it will be that information known only to one or two members will be introduced or discussed. For instance, if there are eight people in the meeting, then for $p$ = 30 percent, as before, the probability that any shared item will be discussed is 88 percent.

For a meeting of two, however, only the graphs for $n$ = 1 and $n$ = 2 are relevant, and here the disparity is not as great. For example, if we take $p$ = 30 percent again ($p$ = 0.3 in mathematical form used in the graph), then the probability that a particular item already known to both persons will be introduced into the conversation is 51 percent (0.51), compared with the probability of 30 percent (0.3) that a particular item known only to one will be introduced.

Thus, if the aim of a meeting is to exchange information and discuss new ideas, then the statistical considerations argue strongly in favor of small meetings, preferably just two or three persons. The argument becomes stronger when you also take account of the psychological and group dynamics factors: It is psychologically easier to introduce an item of information into a small meeting than into a larger group, and it is easier to ensure that two or three people actually discuss a new piece of information than it is for a larger group.

In one study,* three- and six-person groups of university students were given descriptions of three hypothetical candidates for student body president and asked to select the best candidate. The descriptions were constructed so that some information about the candidates was read by everybody in the group, whereas other information was read by only one member. Overall, the groups discussed 45 percent of the shared information but only 18 percent of the unshared information. The difference between the percentages was greater for the six-person groups than for the three-person groups. (In fact, for the larger groups, most of the unshared information was not even mentioned during the discussion, let alone discussed.)

*Stasser, Taylor, & Hanna (1989).

# How can we force ourselves to tread new ground?

There are several ways that organizers of meetings can impose structure to try to counter the tendency for a group to focus on previously shared information. Participants can be asked to write down, prior to the meeting, those points they think should be raised. The meeting could follow a round-robin format, where each participant is given an opportunity to present his or her ideas. Another possibility is for the chair of the meeting to begin by stating that a major goal is for individuals to bring new information to the group's attention and to repeat this instruction from time to time during the course of the meeting. The systematic use of common information artifacts such as flipcharts or whiteboards, discussed earlier, has also been shown (Taylor 1990) to increase the number of ideas incorporated into the final decision.

You can also expect a greater exchange of new information in a meeting when those present each have clearly defined—and mutually understood—areas of expertise: say the designer, the construction specialist, the sales manager, the accountant, the lawyer, and so on.

A less formal version of the "areas of expertise" approach can arise with teams that are used to working together. Over time, the team members learn what kinds of contributions each of them can usefully make, and they can then probe and ask for specific kinds of input at appropriate points of the discussion. To some extent, this approach can be engineered by having the group work in advance on suitably chosen artificial problems as part of an initial team-building exercise.

Stasser (1992) suggest that a further benefit can be gained by formulating the goal of the meeting in terms of finding a correct or best solution as opposed to making a judgment. He points out that when participants conceive their task as making a judgment, they are likely to be strongly influenced by the pool of previously shared information. They selectively choose the information they have that supports the judgment that follows from that shared pool, denigrating information that runs counter to that judgment. If, on the other hand, they see their tasks as searching through all the available evidence to find a correct or optimal solution, they are more likely to want to examine all the information, shared or not.

## SUMMARY

It is known from observation that the more participants there are in a meeting, the greater the likelihood that the group will spend most of the time discussing information already known to the majority, if not all, of the participants.

Simple mathematical considerations indicate why this is the case. According to both experimental observations and mathematics, the "going over familiar ground" phenomenon is very hard to avoid with meetings that involve four or more people.

Among the various ways that meeting organizers can try to counter this trend are:

1. Get participants to submit in advance the points they wish to raise.

2. Adopt a round-robin format where each person in turn is asked to contribute something new.

3. List each new item introduced on a flipchart or a whiteboard.

4. Constantly remind the participants that the aim is to examine new information or ideas.

5. Cast the task at hand in an open-ended fashion as one of examining all the options rather than of making a judgment or arriving at a decision.

6. Ensure that everyone in the group has a clearly defined and clearly understood area of expertise.

7. Build up the team over time, so everyone becomes familiar with one another's areas of expertise and with their strengths and weaknesses.

# 13

# Lambs to the Slaughter

## DEAR MUM AND DAD

So far, we have regarded the contexts in which people converse or act as pools of information to be drawn upon (and possibly exchanged). But there are, of course, other aspects to context besides informational. In this chapter and the two that follow, we consider some of the other ways that context influences the way we act.

I'll begin by reproducing a letter that the former Dean of Student Advising at the college where I work used to read to the assembled parents of the class of entering freshmen students at the start of each new academic year. He began by saying that the letter had been sent home to her parents by a young woman student who had entered the previous year. Here is the letter the dean read:

*Dear Mom and Dad,*

*I am sorry I have not been in touch with you for several weeks now. I think I should explain right away what has been going on. About a month ago, at a party in a frat house, I am afraid I had far too much to drink, and fell down the stairs. I broke both legs and suffered a severe concussion. When I regained consciousness the next morning, I learned that a member of the frat house had brought me to the hospital in the back of his pickup truck. Of course, he too was drunk, so he shouldn't have driven really, but luckily nothing happened.*

*I am still in a wheelchair, of course, but the casts should come off any day now, and the doctors says that pretty soon I should be able to walk just as if nothing had happen. Also, the headaches are starting to go away now, and I am starting to sleep through the entire night. The left side of my face continues to twitch a lot, but the doctors are hopeful that it will also go away in time.*

*The good thing about all of this is that Chip, the man who drove me to the hospital, and I have fallen in love and decided to get married as soon as I am out of the wheelchair. We've already filled in all the forms to withdraw from college. Chip's father owns a fishing tackle and bait store in Idaho, and has said that we can go and live there. Chip will take over the store, and I will help him.*

*Dad, I'm dying for you to meet Chip. Please don't be put off by his tattoos or by the rings through his ears and his nose. He's really very sweet, and I'm sure that once you become acquainted, you'll like each other.*

*Please forgive me for not writing sooner about all of this, but I knew you would worry, and since I had Chip to look after me, I thought it better to wait until I was getting better.*

> *Your loving daughter,*
> *Kathy*

*P.S. Don't panic. None of the above is true. There was no accident. I am just as healthy as when you last saw me. There is no Chip, and I am not planning to get married, or give up college. But I did fail freshman chemistry and I got a C in math, and I wanted you to see these results in a proper perspective when I told you.*

Like me, if you were one of Kathy's parents, you would probably not appreciate receiving such a letter. Kathy appears to score as poorly when it comes to sensitivity as she did on her chemistry and math exams. But for her appreciation of human psychology, I have to give her an *A*.

Psychologists have known for years that all of us, no matter how educated and sophisticated we regard ourselves, are subject to the effects of prior conditioning. As we have noted already in this book, everything we do occurs in a context, and is influenced by that context. We can't avoid it. In the hands of a skillful persuader, like the man who bought the farm from the elderly couple, described in

Chapter 10, most of us are virtually helpless. All the persuader has to do is establish the right context, and before we know it, we have been led exactly where the persuader wanted us to go. Like lambs to the slaughter.

In this chapter I'll try to show you just how significant is the role played by context in many—indeed, most—of our everyday activities. If you are still not convinced of the need to have a "science of context," then this chapter should provide you with the final evidence.

## GOOD MORNING, MAY I HELP YOU?

No one knows the powerful influence that context can exert more than a successful salesperson. Think back to the last time you went shopping for clothes in a large store. Did the salesperson show you the suits first and the less expensive items, such as the shorts and sweaters, later? If he or she did not, they are unlikely to last long in sales. As any successful salesperson knows, and as has been confirmed in numerous studies, we are far more likely to shell out $50 on a sweater just after we have spent $400 on a suit. In the context of having just bought a $400 item of clothing, a price tag of $50 seems paltry. It doesn't matter that we might have had no intention of buying a sweater in the first place. Compared to $400, spending $50 seems insignificant.

On the other hand, if the salesperson had spent the first fifteen minutes showing you a range of sweaters in the $40 to $70 range, the cost of a suit at $400 will seem very high, and you may well refuse even to try it on.

The point is, your first purchase will provide a context for your subsequent transaction(s). Skillful salespeople know this, and will generally begin by showing you items that they know or suspect are more expensive than you are willing or able to afford. They know you are unlikely to buy something at that price. They are in fact not trying to sell you any of those expensive items. They are simply establishing a context—a price context—within which the items they do want to sell you will appear much more affordable. It may well be that the price of those "cheaper" items is still more than you originally intended to pay. Compared to the first range of items you were shown, however, they are less expensive. A price that would have seemed too high before you entered the store suddenly appears acceptable. What the salesperson has done is change the context in which the transaction takes place—exactly as did the negotiator who bought the New England farm from the elderly couple.

Perhaps the most prototypical of sales folk, and certainly the most maligned, is the secondhand car salesman. Successful second-hand car salesmen (and they are almost all male, more often than not called Jim or Ted) will always start out by showing you a car far more expensive than you want to pay. You walk onto the lot saying you are looking for something reliable that costs, say, between $4,000 and $5,000. "I think I have just what you are looking for," the salesman beams. The car he takes you to carries a price tag of $5,999. If he strikes it really lucky, you will automatically regard $5,999 as being "in the five-thousand dollar range," even though it's just $1 short of $6,000, and he may persuade you to "get in and sit behind the wheel." But even if you have your wits about you, and protest that a price of almost $6,000 is too steep for you, the sales-man has already achieved his main aim. Anything he shows you from now on that costs significantly less than $6,000 is going to appear much more acceptable. After all, he might point out, proud-ly displaying a model priced at $5,299, this car will save you $700. Save you? You never had any intention of paying $6,000 for a car. How can the salesman be saving you money on a purchase you nei-ther made nor wanted to make? No matter. We're not talking logic here, we're dealing with psychology.*

Though they tend not to come in for so much criticism, dealers in new automobiles are just as skillful at establishing contexts in which we can be parted with far more cash than we intended. That $16,899 price sticker on that new car you fancy is the price for the "basic" model, the one without the electric windows, the electrically operated front seats, the air conditioning, the CD player, the tinted windows, the special trim, and so on. A couple of hundred dollars here, three hundred there, the cost of each additional feature seems so reasonable when you have already committed to spend almost $17,000. After all, you are just adding a few hundreds on top of many thousands. "Have you considered the supercharged model?" the salesman asks. "I think you'll find it's a lot more fun to drive." And so the transaction goes. In the end, you find yourself driving

---

*Actually, as I point out in my book *Goodbye, Descartes*, there is a logic here, but it's a context-dependent logic. In fact, it is the need to take context into account that led to my title, *Goodbye, Descartes*. Cartesian logic, the kind that usually goes under the name "logic," ignores context.

away in a new car for which you have just paid a cool $20,000. It may be worth it. You may indeed feel very pleased with your new car. But it was that newspaper ad that featured the $16,899 figure that first persuaded you to enter the salesroom. When you set out that morning, your intention was to buy a $17,000 car. That was the vehicle you wanted. Are you an indecisive wimp who doesn't know your own mind? No. You are simply a normal human being who is subject to the generally very powerful influence of context on your actions.

## WELL, IT DOES NEED A BIT OF WORK

Real estate agents often use an approach similar to car salesmen. Let's say that you set out to buy a house in the $150,000 to $200,000 range. The agent will do one of two things, maybe both. Either you will find that the first two or three houses you are shown all cost upwards of $240,000. Or else you'll find yourselves looking round two or three houses that are definitely in your price range but are absolute dumps you would never dream of buying. "Well, this one does need a bit or work," the agent will admit as you pull up in the driveway. "You're telling me," you say to yourself as you step out of the agent's car. Surely she doesn't expect me to buy this, you think. And you are absolutely right. She knows full well you would not touch this house with a 10-foot pole. She's not trying to sell you this house. She's establishing the right context for showing you the houses that she thinks you probably will buy. "I know it's a bit more than you wanted to pay," she says later, standing in the living room of an attractive property listed at $235,000. "But it's very sound and in move-in condition. The builder is local, and has a great reputation for solid, old-fashioned construction methods."

When you started out that morning, you were determined not to go a penny over $200,000. Notice that the realtor never made any attempt to persuade you to change that figure. She simply changed your context. As an experienced salesperson, she knows that the phrase "the maximum amount I want to pay" almost never denotes a fixed sum of money. It's context-dependent. (Linguists would say that the phrase is "indexical," which is their technical term for any word or phrase whose exact meaning depends on the circumstances in which it is used.)

Of course, it is possible to resist context-changing tactics. Some people seem to have a high level of such resistance, but for most of us it takes considerable effort. In our natural state, we are far more susceptible to the influences of context than we might care to admit.

Let me give one more example, this time based on studies carried out at a number of universities. Our judgment of physical attractiveness depends on the context in which we make it. In one study, college students were asked to rate the attractiveness of members of the opposite sex, based on photographs they were shown. The photos showed individuals that most people would rate as "average looking." When the subjects were shown the photographs after spending a few minutes looking through various popular show-biz magazines, they rated the individuals in the photographs as less attractive than they did when they had not been so primed. The act of looking at the magazine photos of the unusually attractive movie stars established a context within which the photographs of the "ordinary" people seemed far less attractive than would otherwise be the case.

In another study, male students were asked to rate the attractiveness of a "blind date," whose photograph they were shown. Their ratings were far lower when they were shown the photograph after watching an episode of the then-popular television series *Charlie's Angels*, in which the heroines are played by three particularly attractive female actresses. Same students, comparable photographs chosen randomly, but a different context.

## COMMITMENT COUNTS

Have you heard the one about the unsuccessful Japanese kamikaze pilot in World War II? He survived nineteen missions. As he said later, "I had all the right training and skills, I just didn't have the commitment."

Commitment is a powerful factor in almost everything we do. Just how powerful it can be was made abundantly clear by some startling results that were published in the *Journal of Personality and Social Psychology* in 1966. The authors were the research psychologists Jonathan Freedman and Scott Fraser.

In their paper, Freedman and Fraser reported the results of an experiment in which a researcher, posing as a community volunteer, had made door-to-door visits in a residential Californian neighborhood. Householders were asked individually if they would consent to a public-service billboard being erected on their front lawns. They were each shown a photograph of what it would look like. The photograph showed a pleasant, attractive, suburban home, almost completely obscured by a large and badly made sign that read DRIVE CAREFULLY. No one in their right mind would consent to despoiling their home in this manner, you might think. And indeed, 83 percent of one group tested refused to allow such a sign to be installed.

However, a second group responded very differently, with a staggering 76 percent saying yes to the request. What was the difference between the two groups? In terms of their selection, there was no difference at all. The researchers had divided the entire test population into two groups in an entirely random fashion, alike in all observable respects. The approach to the members of the two groups had been exactly the same: same request, same photograph.

What was different was that the members of the second group had been visited two weeks previously by a different "volunteer worker," who had come to their doors to ask them to display a small (3-inch-square) sign that read BE A SAFE DRIVER. It was such a small request that most people complied willingly, without giving the matter much thought.

And indeed, it was a small, seemingly insignificant request. But its consequences were anything but insignificant. By accepting and displaying the sign, the homeowners had accepted the role of "a public-spirited citizen who supports safe driving." [CAUTION: *A new context has just been established!*] As a result, two weeks later some three-quarters of them were willing to comply with a truly preposterous request.

In fact, the Freedman and Fraser paper report on an even more startling finding. The two researchers carried out a slightly modified procedure on a third group of homeowners. Like the members of the second group, the homeowners in this third group were also visited two weeks prior to the billboard request. But for this group, the request was simply that they sign a petition calling to "keep California beautiful." Well, ask yourself, wouldn't you sign? Who can possibly object? And what possible difference can it make to you if you sign?

Well, for many people in this third group, simply signing the petition made a *huge* difference. Roughly half of them subsequently agreed to having the large billboard erected on their front lawns! The explanation offered by Freedman and Fraser for the behavior of the members of this third group of subjects was that the seemingly trivial act of signing a petition to keep the state beautiful had changed their self-image. They now saw themselves as public-spirited citizens, prepared to do their bit for the common good. They had acquired a commitment, and that commitment provided a powerful context for further actions.

The evidence from the examples discussed in this chapter is clear: To ignore context is to give up any hope of properly understanding human behavior.

## SUMMARY

Context provides a very powerful influence on most of our actions.

Salespeople will frequently start out by showing a customer a more expensive item, to establish a context in which the desired item appears less expensive.

People's judgments of physical beauty are conditioned by the images they have last been subjected to; subjects who looked at photographs of members of the opposite sex viewed them as less attractive in the context of having just looked at photographs of attractive movie stars.

A very minor act can cause someone to make a commitment to a cause. But no matter how minor the initial act is, the commitment that results can exert a significant influence on the individual's later actions and decisions. A minor initial act of seemingly no significance can lead people to make decisions highly detrimental to themselves—decisions they would otherwise not have made.

# 14

# A Fine Power

---------------------------------------------------------------

## THE FIELD ENGINEER'S HIGHLY CULTURED BEHAVIOR

Another contextual influence on human behavior, that can be every bit as powerful as the ones described in the previous chapter, is that of culture.

As a first illustration, consider the PRF data from the computer company that Rosenberg and I examined (described in Chapter 5). One of the factors that caused so much difficulty, both for the experts in the computer company who had to obtain information from the Problem Report Forms (PRFs) and for Rosenberg and myself as analysts, was the strong influence of social and cultural factors in the way the documents were completed. Despite the fact that the PRF was limited entirely to the domain of one company's computer systems, and despite the fact that each PRF was filled in by technically trained experts (customer service personnel and field engineers), in many cases obtaining the desired information from the PRF required a familiarity with social and cultural issues.

For example, interpretation of many of the PRFs required a knowledge of the working practices of field engineers. Though designed as a means to obtain information about faulty computer systems, the PRF could also be used to track the performance of field engineers. This fact was not lost on the engineers themselves, and indeed many of them viewed the document primarily in this way. Since computer engineers tend to be fairly individualistic, this potential use of the PRFs caused considerable tension and influenced the way the engineers completed the document. The result was often poor information about the faulty computer system.

One PRF that Rosenberg and I examined concerned a computer system for which a printer fault had been reported. The action reported by the field engineer was simply that he had contacted the on-site operator. This report was incomprehensible to everyone except another engineer. The problem had turned out to be a simple paper jam. Annoyed that he had been called out for such a trivial problem, the engineer had refused to remove the jam and had insisted that the operator do it. He even refused to say what the problem had been! It was a classic case of job demarcation.

With social and cultural factors playing a role in such a constrained and technical domain as computer fault reporting, it should come as no surprise to discover that in less technical environments they can play an even greater role. The plain fact is, we are social creatures and we live in a culture, and that culture is an ever-present part of the environment—the background situation—in which we communicate.

## OUT OF THE MOUTHS OF BABES

Just how pervasive social and cultural influences are was brought home forcefully by the late Harvey Sacks in some groundbreaking work carried out in the late 1960s and early 1970s. Trained as a sociolinguist, Sacks took a very close look at simple, everyday uses of ordinary language, including conversations. In one paper, which was initially presented in a lecture and published later, after his untimely death, he analyzed the way small children used language.*

One of the main examples Sacks examined in that paper was the following two-sentence opening of a story told by a three-and-a-half-year-old child:

*The baby cried. The mommy picked it up.*

As Sacks observes, no native speaker of English has any trouble understanding what the child has said. You don't have to stop and think about it. Sacks then asks you whose mother is being referred to. "The baby's mother, of course," is the automatic reply. But there

---

*The paper, which is now regarded as a classic, was called "On the Analyzability of Stories by Children," and it was published in the 1972 volume *Directions in Sociolinguistics, The Ethnography of Communication*, edited by John Gumpertz and Del Hymes (pp. 325-345).

is nothing in the data that tells us that. It comes from social and cultural knowledge.

When did the mother pick up the baby? Sacks asks. "When she heard the crying," you respond. Again, this is not in the data; it comes from your social and cultural knowledge.

Why did the mother pick up the baby? "Because it cried, in order to comfort it." Once more, this automatic assumption comes from social and cultural knowledge.

Sacks has more to say in the same vein about this particular example, and about other examples. Social and cultural influences on our use of language to communicate is not a peripheral thing, it is central, he argues, summarizing his observations with these words:

> *My reason for having gone through the observations I have so far made was to give you some sense . . . of the fine power of a culture. It does not, so to speak, merely fill brains in roughly the same way, it fills them so that they are alike in fine detail. The sentences we are considering are after all rather minor, and yet all of you, or many of you, hear just what I said you heard, and many of us are quite unacquainted with each other. I am, then, dealing with something real and something finely powerful.* [Sacks, 1972]

In fact, it is misleading to talk, as I did above, about "making assumptions" and basing them "on social and cultural knowledge." Sacks's point is not that we make assumptions when we understand everyday language, let alone that we base those assumptions on knowledge we already have. Rather, he says, *the very way we hear what is said* is influenced by our social and cultural background. The words that we hear enter our minds already conditioned by our social and cultural environment. This is what he means when he speaks of "the fine power of a culture."

The same is true when we speak. We may feel that we are free to choose whatever words we wish, but the evidence to the contrary is overwhelming. For example, each of the terms "police officer," "policeman," "cop," and "pig" can be used to refer to a policeman, and in that sense they are equivalent. But, depending on the circumstances, use of only one of these four terms will be appropriate, and use of any of the others in an inappropriate situation will cause the listener to question what the speaker really means.

As an exercise, the next time you are in a conversation, remember some of the words you use, and then, after the conversation is done, imagine what would have happened if you replaced each of those words by another one having the same literal meaning. For example, what would have happened if you replaced "woman" by any one of "lady," "female," "gal," "broad," or "Shiela" (an Australian term)? Or compare the different "meanings" of the phrases "the portly lady" and "the fat woman."

## WHY BOTHER WITH POLITE CONVERSATION?

Not only does our social and cultural environment influence—extensively and to a precise degree—the words we use and the way we understand words we hear and read, it also affects the way we structure our interactions with one another—the way we say things to one another. Being aware of these cultural influences—and why they arise—can be extremely valuable when it comes to that important meeting where you have to get your point across.

Why, for instance, do we typically say "Do you have the time?" when our goal is to be told the time? Literally speaking, the question "Do you have the time?" requires a reply of "yes" or "no," but we all know that to respond in such a way would be both inappropriate and very impolite.

Why do we say "Can you pass the salt?" when our intention is for the person to actually pass the salt? After all, we know that they can!

There are several different explanations for such behavior, though the distinctions between them may seem minor to anyone but a sociologist. The explanation I shall give depends on a concept known as "social equity." Roughly speaking, social equity is a balance—an equilibrium—that members of a society strive to maintain. Why? Because that is part of our nature as social beings.

As we observed already, any kind of conversation has to be entered into as a joint act—the two (or more) people involved have to collaborate, if only to a minimal degree. Except in situations in which one person has a clearly understood position of authority over the other, the participants in a conversation will (usually subconsciously) go to some lengths to ensure that nothing is said that would imply authority, and thereby be taken as aggressive. Instructing someone to tell you the time or to pass you the salt implies that the speaker has authority over the listener. To avoid this, society has developed a ritual alternative, the "Can you . . . ?" question. This alternative nominally puts the speaker into a "supplicant" position, with the listener having the freedom to say yes or no.

In theory, there is nothing to prevent you from trying to dispense altogether with such "niceties." But if you do, you should be prepared to suffer the consequences. Research has shown just how well-established are these pro forma requests and how uncomfortable most people feel when confronted with a more direct mode of speech.

People who have a background in science are often more likely to want to dispense with some of the rituals we use to interact, feeling that the important thing is to "get the information across." However, while it is true that the "laws of conversation" are not exactly the same as the laws of physics, in that we cannot break the latter but are free to break the former, the similarities are greater than many people realize, and socially based "laws" are not easily broken, at least not without consequences. In both cases our behavior is affected by those laws, and in neither case do you need to be aware of the laws to be affected by them. Step off the top of a tall building and you will suffer the effects of the law of gravity. Step outside the procedures governed by the laws of conversation and you are likely to suffer an equally painful fall.

Because of the undesirable consequences that can follow from a breach of the social norms that maintain social equity, smart people often employ a variety of techniques to avoid such a breach. For example, suppose you are faced with having to broach a topic with someone whom you know does not want to discuss it—perhaps because she finds it painful or else feels defensive about it. In such a situation, a particularly helpful device to use is the "Columbo technique," named after the long-running television detective series *Columbo*.

Inspector Columbo is a master investigator. His approach when questioning a suspect is, first of all, to adopt an "I'm just an ordinary guy doing his job" approach, avoiding any suggestion of the authority he has as a police officer. Second, he spends most of the conversation talking either about general matters (often his wife) or about issues in the case that do not put the other person on the defensive or make him or her angry. Then, right at the end, when the conversation appears to be over and he is preparing to leave, he brings up the key issue in a feigned absent-minded manner, usually prefaced by a remark such as "Oh, I almost forgot, there's just one more thing I wanted to ask."

Physicians in general practice report that patients often behave in a similar way. They spend most of the consultation talking about general aches and pains and about their family's health, and only when they are preparing to get up and leave do they raise the matter that prompted them to come in the first place. "Oh, there is one more thing, doctor," they say as they appear to prepare to stand up and leave.

In Columbo's case, the fictional detective uses the first part of the conversation purely to prepare the way for the real question—to establish a conversational context likely to ensure he gets an answer. Similarly, the patient visiting the physician uses the initial conversation to set the scene for bringing up the real reason for the visit.

In acting in the manner they do, both Columbo and the patient are operating within the framework of human culture. In the next chapter, we'll look at some examples that show how the influence of cultural and social factors on information flow can have major consequences for entire industries and even whole nations.

---

## SUMMARY

Social and cultural factors influence the way we understand the words we hear and the choice of the words we speak. Their role is significant and pervasive and cannot be ignored.

Likewise, social and cultural factors affect the way we interact. The need we feel to maintain social equity forces us to structure our commands, requests, and replies in particular ways. Our purpose in entering a particular conversation might indeed be to give, receive, or exchange information. But the conversation itself is a social activity, governed by the unspoken rules of conversation. In general, it is far more effective to work within those rules.

The Columbo technique can be particularly effective in establishing a discussion about an issue that at least one of the parties finds uncomfortable for some reason.

# 15

# Culture at Work

----------------------------------------

## WHERE IS THE REAL INFORMATION NETWORK?

Even when you have a reasonable understanding of what information is—how it is represented and how it is transmitted—managing it efficiently can still be difficult. When things go wrong, social and cultural factors are often found to be the problem. Often, they are the most difficult issues to address. Here is one example of the unforeseen problems that can arise from what seems, on the face of it, a straightforward exercise in improved information management.

A few years ago, the U.S. Navy sought to update and improve their system for keeping track of spare parts. Prior to the mid-1980s, each ship kept all its records on index cards. This practice made it difficult to keep track of the stock of spares at fleet level, and ships often found themselves with a shortage of some needed items and a surplus of others. The inefficiencies were estimated to cost the taxpayers hundreds of millions of dollars a year.

The obvious solution was to install record-keeping computers on all the ships and link them all to the shore-based fleet logistics planners. However, when this was done, not only did the new method meet with strong opposition from the ships' officers but also the fleet's spares readiness deteriorated significantly.

Here is what happened. There were in fact two spare parts distribution networks, the official one and an unofficial one. The unofficial one, which was the one most often used to obtain important items, took advantage of the slack in accountability in the official one. Ships' officers developed their own underground network, based not on rank or mandated procedures but on individual trust. Over the years, those networks had grown into a large and effective way to obtain spares.

By increasing the efficiency of the official network, the introduction of the new computer systems made it harder for the unofficial network to operate. Greater control was put into the hands of the shore-based logistics personnel. By and large, the naval officers had little faith in their shore-based colleagues; they preferred to deal with their own kind. Thus the situation with the spare parts actually got worse.

That was not the only unintended consequence of what had been intended to be a move toward greater efficiency. The adaptability and ingenuity shown by officers, together with their trust in one another, on which the unofficial spare parts network depended, are precisely the qualities that are important when it comes to combat. The unofficial spare parts network actually played a significant role in maintaining that trust. Thus, reducing its effectiveness by the introduction of the new computer inventory system had the overall effect of reducing overall potential combat effectiveness—though fortunately this was never put to the test in a battle situation.

The story of the navy supply network provides us with a number of cautionary lessons. First, an official information channel may not be the one actually used. Culture very often takes over from mandated procedure. Second, any information channel may evolve to do something other than transmit the information for which it was designed. In the end, that unintended effect might turn out to be just as important as the original, designed purpose.

## A LESSON FROM JAPAN

In the navy example considered above, the difficulty with automating the spare parts information system was that it automated the "wrong" information system. The official network was automated, whereas the system in actual use was an unofficial, socially established network (which of course could not be automated). The unintended side effect of the automation was to damage an important social network that helped to support the important culture of naval officers working as self-reliant teams.

The U.S. automobile industry provides an example in which automation was deliberately accompanied by a change in the working culture, with highly successful results.

Following the rapid growth in sales of Japanese automobiles in the United States during the 1980s, considerable effort was put into trying to determine the causes of the relative inefficiencies in the U.S. automobile industry. Automation was part of the answer, but only part. Cultural issues played a significant role as well. American automobile manufacturers not only automated their plants but also changed their management methods and the way they structured their work force.

Today, the U.S. automobile industry is in some aspects one of the most efficient in the world, and U.S.-produced autos are both good value and extremely reliable. Unfortunately, the dramatic improvements in manufacturing have had less impact on the retail market than they deserved. Among the American public, it is still widely— though erroneously—believed that American cars are not as good value as Japanese or European cars, nor as reliable. The American automobile manufacturers have so far been unable to change the cultural beliefs of a large segment of their potential market.

In terms of changing manufacturing, the case of the U.S. auto industry is in large part a success story. Getting rid of similar inefficiencies in the defense manufacturing industry has proved to be more difficult. It takes the United States far longer to design and manufacture, say, a new missile system than it does Israel. This, despite the enormous wealth and technical expertise in the United States, a country which is the clear world leader in the development of new computer technology and software.

The problem is almost certainly cultural. The structure of the defense supply industry requires the interaction of two very different cultures. First, there is the military. In the United States (but not in Israel) the armed forces have a highly structured, hierarchical culture with many formal procedures. The other culture is that of the high-tech industry. The world over, high-tech companies tend to have a very flat management structure and a very free-and-easy way of doing business. In the United States, the two cultures are so very different that it has so far proven impossible to find the efficiencies being sought in the development of new missile systems. In contrast, because of Israel's small size and the nature of its society, with all citizens serving in the military on a regular basis, the Israeli military has a culture far closer to that of the high-tech people with whom it must work to develop new weapons systems.

For some years now, a friend of mine called Ted Goranson has headed a loosely knit research group (The Agility Forum) that has received government sponsorship to look for ways to overcome the problems facing the defense supply industry. As part of that research, Ted has been looking at a small number of cases of very efficient industries, to see what makes them work. Culture is almost always the key. One of the more interesting examples he has found is the present-day movie industry.

Despite those occasional headlines about exorbitantly paid movie stars holding up production and causing massive cost over-runs, when viewed as a collaborative enterprise, the movie industry is extremely efficient.

Movies are no longer made by monolithic motion picture companies. These days, to make a movie, a large number of people, all highly skilled in different ways, come together for the time it takes to get the thing made, and then disband. For the most part, the whole process takes place without any oversight from someone "at the top."

Most of the hiring and subcontracting is done on an informal basis, through personal contact and personal referral. Though contracts are drawn up, what holds the enterprise together is, for the most part, tradition and culture. That's why the movie star Kim Bassinger was bankrupted a few years ago when she reneged on a simple, verbal agreement to make a particular movie. Also relevant is the fact that the legal system backing up the contracts is based on case law, not code. This allows for far more flexibility when new situations arise. Moreover, the movie industry has always proved particularly quick to adopt new technology.

How does this system work, and what would it take to emulate it in other industries? Can a successful working culture be engineered? Goranson investigated the issue, and here is what he found out.

Prior to the start of the movie industry, Hollywood was an oil town, and the movie people adopted the working practices and the legal infrastructure already in place. In turn, those working practices and the associated legal infrastructure came from Pennsylvania, the home of the oil business. Pennsylvania inherited them from the previous oil industry: whaling.

## FROM WHALING TO MOVIES

For many years, whale oil was a highly significant product in the economy. By providing candles that were much brighter and less smoky than the tallow candles that preceded them, the production of whale oil led to a change of lifestyle to one where many activities could be carried out at night. Moreover, the fine lubricating oil that comes from the head of the sperm whale played a key role in the ability to mass-produce the pocket watch, another development that changed the working day. (In fact, lubricating oil from sperm whales was used in high-precision machinery up until the 1970s. For instance, it was used in the gyroscopes in the Apollo mission to the moon.) Thus, whaling was an enabling industry that helped to drive many other industries, in this respect not unlike today's microchip industry.

The whales from which the oil was extracted were found mainly in the Pacific and Indian oceans. And yet, for several decades, the whaling industry was dominated by two small towns far away in New England: Nantucket and New Bedford. The whaling expeditions launched from these two towns produced 90 percent of the world's whale oil.

What was the secret that gave these two small communities such dominance? The most significant factor seems to be their size: They were small communities, each with a single, strong culture. Their legal system was based on case law (managed by the court in Boston), not legal code. (In situation-theoretic terms, their legal system depended on types, not rules.) Agreements were made and sustained largely on trust. Crews were assembled through personal contact and personal recommendation.

In many ways, it was simply a historic accident that the two communities of Nantucket and New Bedford were the first to gain dominance in the whale oil industry. But once they achieved dominance, it was largely because of cultural factors that they kept it, *not* expertise in whaling. Indeed, when the California gold rush began, some whaling teams abandoned whaling and rushed instead to the gold fields. Once there, they were highly successful, again as a result of their collaborative working practices based on cultural bonds. Each member of the team simply focused on a different task. Acquiring a new skill is relatively easy, whereas building an effective working culture can be extremely difficult and time consuming.

A classic example of how people can change their specific tasks but continue to serve the community is provided by sailmakers. Each whaling party had its own sailmaker. One such was a man called Levi Strauss. Finding little demand for sails in the Californian gold fields, Strauss started to make durable pants for the miners out of the canvas and rivets previously used to make sails. Blue jeans were born, with Levi's own brand being the first. The rest, as they say, is history.

The key factor, then, that maintained the dominance of Nantucket and New Bedford in the whale oil industry was their community ties, which gave rise to well-coordinated teams, with agreements being based largely on trust, backed up by the highly flexible case law system. This cultural structure was far more important than any technical expertise.

Indeed, specific expertise can be acquired with comparative ease, whenever new circumstances demand change. In contrast, culture tends to endure and to resist change. Thus, the cultural structure of the New England whaling communities was passed on to the petroleum oil industry when petroleum oil started to take over from whale oil, and then on to the movie industry when film making sprang up in the (then small!) oil town of Los Angeles. *The industries kept changing; the working culture remained largely the same.*

Another illustration of "the fine power of a culture" was provided by the movie industry itself, in the case of the 1995 film *Waterworld.*

## WATERWORLD

Directed by movie superstar Kevin Costner, who also starred in the film, *Waterworld* was a financial disaster until the losses were eventually recouped through video sales. Interestingly, the huge budget overrun during production can be traced to a single decision that ignored cultural issues. Here is what happened.

An unusual feature of this film was that it was almost all filmed on floating sets, two of which were large and complex. As is generally the case, most of the sets, including the two large ones, had to be constructed from scratch. The question was, who was going to build them? If the usual set construction crews were to do the work, they would have to be sent to school to learn the marine engineering and the safety skills required for the construction of floating sets. When the question arose early in the planning for the movie, there was plenty of time to do this, and it was costed out at about $3 million.

The other alternative was to go outside the normal Hollywood community and subcontract to an established marine engineering company which already knew and understood all of the engineering and safety issues.

The filmmakers chose the former approach. They preferred to work with people from the community and the culture they were familiar with, people with whom they knew how to work. However, the Japanese conglomerate that owned the studio vetoed this route as too expensive, and they insisted that the work be carried out by an established marine engineering company.

There were problems right from the start. The construction people need to be familiar with the special demands of movie making, where the artistic effect is all important. In particular, the construction team should be able to *interpret* what the director is asking for. Movie sets are designed to convey a certain "feel" that can be captured by a camera. They must also accommodate the mechanics of camera placement. Moreover, once filming starts, frequent changes to the sets are called for by the director. These are generally conveyed directly to the construction crew, most often verbally, sometimes illustrated with rough sketches. Changes often need to be carried out overnight so the set is ready for the next day's filming. Discussions are carried out using the artistic terminology familiar to movie people.

All of these features of traditional movie set construction were unfamiliar to the marine engineers who were contracted to make the sets. They were used to working sequentially from carefully prepared blueprints. In their world, communication was achieved using engineering terminology. Any changes had to be worked through and evaluated at length. Artistic effect was an alien concept.

The enormous difference in culture between the Hollywood people and the set construction crew caused continual problems. Production costs skyrocketed as numerous small events slowed down the process and compromised the artistic value of the final product. On one occasion, one of the sets accidentally sank. In the end, the conflict between the two cultures caused the production costs to run over budget by $80 million.

The lesson, once again, is that *culture is the key to an efficient working infrastructure; culture endures and resists change but specific expertise can be acquired.* In particular, "re-engineering the corporation" is not nearly as easy as that popular slogan might suggest.

The remainder of this book consists, in large part, of a number of examples to illustrate the above point, and to show how the wise person can turn information into knowledge most efficiently and to the greatest effect.

SUMMARY

Numerous studies of successful and unsuccessful industries have shown the important role played by social and cultural factors: the way people work together and the procedures and infrastructure that support their work. These factors are not easily engineered; more often, they are simply passed on from one generation to the next, as new people are brought into the culture and become familiar with it. In contrast, specific knowledge and skills are, in general, comparatively easy to acquire. Consequently, it is often easier for people to change *what* they do than *how* they do it.

# 16

# Forever on Your Mind

-------------------------------------------------

## GET OVER TO THAT WATER COOLER AND DO SOME WORK!

John, Alice, and Kevin all work for the same large, multinational computer company. John is always to be seen at his desk, pen in hand or fingers over the computer keyboard. He generates many memos and he answers all his email promptly. Alice, on the other hand, can often be seen sitting in a comfortable chair, reading a (work-related) book. You are as likely as not to encounter Kevin chatting to his colleagues in the corridor, at the water cooler, or by the copier.

Which of the three is likely to be the most valuable employee—the most effective knowledge worker? The traditional management view prevalent in North America and Europe—at least until recently— is that it is John. While Alice is "wasting her time" reading and Kevin is "wasting his time" gossiping, John is "hard at work" at his desk. He is the one who will get the key promotions, while Alice and Kevin will be prime candidates to be let go at the first hint of downsizing.

In the days when many office workers performed routine, repetitive tasks, the traditional view made sense. For today's knowledge workers, however, it is hopelessly outdated. John may indeed be performing a valuable service to the company. But so too are Alice and Kevin. By reading (work-related) books, Alice is generating new knowledge for the company. By talking regularly to his colleagues, Kevin is ensuring that knowledge flows through the company.

In the knowledge society, the very idea of what constitutes "work" changes radically. In today's leading companies, you are unlikely to find many Johns, Alices, or Kevins. What you will find are people who do all three: desk work (the traditional stuff of "work"), knowledge acquisition and creation (an important sounding way to describe reading, watching a training video, or attending a workshop), and knowledge transfer (an equally important sounding term for gossiping).

Many managers, brought up in an earlier age, have yet to fully grasp that change. Take the case of John Akers, who was chairman of IBM at the time when companies were beginning to phase out their reliance on the large mainframe computers that were the wellspring of IBM's business. IBM's hitherto loyal customers were replacing their mainframes by the smaller minicomputers built by the upstart Digital Equipment Corporation and the microcomputers that had started to emerge from garage workshops in Palo Alto. In a company memo that has since become famous among management scientists, Akers demanded that his employees should stop wasting time at the water cooler and get back to work.

What Akers failed to realize was that, although some of the conversation at the water cooler was indeed social chat about home, family, sport, television, politics, and the like, much of it was directly related to work, often importantly so. With the marketplace changing rapidly, IBM's employees—most of them notoriously loyal to the company—were trying to find ways to respond. Those chance meetings were a significant opportunity for people in different parts of the company to exchange ideas—people with different knowledge and skills and with different experience. By eliminating those chance encounters, Akers took away one of the major mechanisms whereby knowledge flowed through the company. As a result, IBM's ability to respond to the challenge in the marketplace was undoubtedly hampered.

Today, it would take a manager who had somehow managed to insulate him- or herself against years of endlessly repeated management advice to make the same mistake Akers did. Indeed, a number of companies have gone as far as to *engineer* "water cooler meetings," to make sure they take place at regular intervals. Sometimes, this is done by providing designated "talk rooms" and encouraging—or enforcing—employees to spend a certain amount of time there each day, say twenty minutes, simply talking to other employees, free of any preset agenda. Occasionally, when face-to-face meetings are not possible, technology is used. For example, British Petroleum (BP) uses videoconferencing technology to arrange a weekly, twenty-minute "virtual tea break," where up to twenty people at eight separate locations come together to "chat" about their work. (The tearoom is the British equivalent of the American water cooler, of course.)

## TURNING INFORMATION INTO KNOWLEDGE

We have already seen several examples of how information becomes knowledge. (Indeed, as we have observed, the distinction between information and knowledge is not a clean one, and is in large part a matter of emphasis.) Let's try to sharpen up our understanding of this process.

Going back briefly to our imaginary trio of John, Alice, and Kevin, there is little doubt that John generated the greatest information flow. Sitting long hours at his desk, generating all those memos, and sending out all those emails, his entire day was spent receiving, processing, and transmitting information.

But what was *done* with all that information? John himself seemed to do little with it other than to process it in one way or another. Depending on his job, that may or may not have been valuable to the company. But somewhere in the company, someone must do something with at least some of the information that is shipped around.

That's where knowledge comes in, as opposed to "mere" information. But what does it take to turn information into knowledge? And what is required to transfer information versus what is required to transfer knowledge? As we shall see, the difference is significant.

Let's start by recalling what we said about knowledge in Chapter 1: Knowledge results when a person internalizes information to the degree that he or she can make use of it. Or, as Davenport and Prusak defined it:

*Knowledge is a fluid mix of framed experiences, values, contextual information, and expert insight that provides a framework for evaluating and incorporating new experiences and information. It originates and is applied in the minds of knowers. In organizations, it often becomes embedded not only in documents or repositories but also in organizational routines, processes, practices, and norms.*

In the light of our study so far, we are now in a better position to appreciate what these descriptions say.

Though information only arises as a result of human-generated constraints (or constraints determined by some other species), it is somehow outside of those humans (or other agents), and independent of them. As such, it is both reasonable and useful to view information as some kind of "substance" that exists in the public domain. For example, the information that England is a monarchy is available to all and is surely independent of any particular person.

Knowledge, on the other hand, is fundamentally and intrinsically inside people (or possibly members of other species). After all, knowledge implies that there is a *knower*. Davenport and Prusak make essentially the same point, when they write: "The power of knowledge to organize, select, learn, and judge comes from values and beliefs as much as, and probably more than, from information and logic" [Davenport & Prusak, 1998, p. 12].

In this chapter, we shall take a closer look at the distinction between information and knowledge. Then, in the remaining chapters, we shall see what a company can do to manage knowledge to its advantage—to acquire knowledge, to create it, and to transfer it to those people in the organization who need it.

One thing to pay particular attention to as we put knowledge under the microscope is this. Knowledge is information possessed in a form that makes it available for immediate use. As we learned from our initial investigation of information (recall the dinosaur's egg), information depends in a crucial way on context and constraints. Thus, turning information into knowledge requires recognition of—and familiarity with—the relevant contexts and a mastery of the appropriate constraints.

Some of these contexts and constraints are the ones that provide the link between information and its representation (that is, between information and data). These are the kinds of contexts and constraints we examined in the early chapters of the book. But as we saw in the more recent chapters, cultural contexts and psychological and social constraints also play a role—often a crucial one.

## QUALITY COUNTS

I'll start with a metaphor that I find particularly suggestive: that of the fuel used to drive an engine—in this case, an engine that is itself metaphorical.

In the Industrial Society, crude oil was a major source of the power used to drive the engines and power the factories. But before the chemical power in the oil could be unleashed, the raw crude had first to be refined and thereby turned into usable forms such as heating oil and gasoline. Similarly, information is the source of the power that drives the "engines" of the Knowledge Society, but in order to make use of that power, we must first turn the information into a usable form—knowledge.

Turning crude oil into heating oil and gasoline requires a chemical-refining plant. Turning information into knowledge requires a person, someone who, to use Davenport and Prusak's words, has the "power of knowledge to organize, select, learn, and judge," a power that "comes from values and beliefs."

We can take this metaphorical analogy further. Crude oil is measured by volume only. When crude oil is refined, quality becomes important. The process of refining the crude adds value to it: It becomes more useful and at the same time more expensive.

Similarly, information can usefully be measured in volume. Very often, more of it can be a good thing. Which is just as well, given the amount of the stuff that today's information technologies make available to us. But when we refine it to turn it into knowledge, quality counts far more than quantity. When we turn information into knowledge, we add value to it, and make it more expensive. (As Aeschylus observed over two and a half thousand years ago, "Who knows useful things, not many things, is wise.")

Because knowledge is essentially inside people's minds, management of knowledge must be about the management of people. Successful companies, for whom knowledge is a crucial resource, know this full well.

For example, in 1995, IBM acquired the successful software company Lotus. The book valuation of Lotus at the time was $250 million. And yet IBM paid $3.5 billion for the company, fourteen times as much as the book value. Clearly, IBM was not buying "the company" in the traditional meaning of the word. Nor were they buying all the research, development, manufacturing, and product *information* stored in Lotus's file cabinets and computer databases. Rather, the valuable asset that IBM was buying was the knowledge in the minds of the Lotus employees—particularly the expertise of Ray Ozzie, the man who created their most successful office product, Lotus Notes—together with the organizational culture that made that knowledge so effective.

But wait a minute. Hasn't everyone heard of companies that observe and interview their expert employees to capture their knowledge in books, on videotape, and in databases, so that others can acquire that knowledge? Why is it necessary to go to the trouble and expense of acquiring the entire company? Why not negotiate to buy the knowledge in stored form?

The answer is, those "stores of knowledge" don't really encode knowledge; in my view, the phrase is a misnomer. For instance, there is Chrysler's "Engineering Books of Knowledge" project, which attempts to store the knowledge acquired by the design team that develops each new model of automobile. This system undoubtedly provides the user with more than just "data," since Chrysler goes to considerable effort to ensure that the data it collects is given a structure. For example, engineers are encouraged to provide narrative commentaries to accompany any raw data, explaining why a particular design choice was made or why some line of development was abandoned. As a result, what is stored in their "Engineering Books of Knowledge" arguably merits use of the term "information." But I would be reluctant to call the result "knowledge." It still requires a skilled engineer to be able to make use of the information stored in the "Books"—to turn that information into knowledge.

In other words, talking about "storing knowledge" is not entirely accurate—though it's usefully suggestive. A more accurate, but far less catchy, phrase would be "storing information that experts have selected and structured, in order to make it easy for people with similar training to assimilate and thereby convert into knowledge that they can use."

## Eventually, I had to mention Microsoft

Microsoft is another company that understands full well that the only way to acquire knowledge is to acquire the people who have that knowledge.

Newspapers often carry reports of yet another takeover of a small software company by the Redmond, Washington-based software giant. The acquisition is often explained as being motivated by Bill Gates's desire to eliminate all the competition. There may well be something to that. But it would be just as easy for Gates to wipe out a comparatively tiny rival (and these days, most potential rivals to Microsoft are comparatively tiny). All he needs do is develop a competing product and then use Microsoft's huge market dominance to drive the opponent out of business. For a company with

Microsoft's technical resources, it is an easy matter to "reverse-engineer" a similar product having just enough new features to escape the copyright laws.

This is, of course, exactly what Microsoft did when it developed its Windows operating system to provide PCs with a functionality similar to the extremely successful Macintosh. Because of Microsoft's massive financial strength and its dominance of the software market, Windows rapidly overtook and then outstripped the Macintosh operating system as the preferred platform, despite the fact that, by near-universal agreement, the Mac operating system was and remains technically far superior to Windows and much easier to use.

If Microsoft can make PCs with Windows more successful than the Macintosh, then clearly it could use a similar approach with other competitors. But in general it does not. Its preferred strategy is to buy up the company that has developed and manufactured the competing product—lock, stock, and barrel. (U.S. antitrust regulations ruled out the buyout option in the case of Apple, though in 1997 Microsoft bought a share in Apple that was sufficiently large to give it effective control of the company.)

Buying the competitor's company in its entirety usually requires a far greater initial cash outlay on the part of Microsoft than would be required to bring out its own version of the new product. So why doesn't Bill Gates take that route? Because, as the shrewd head of a company that consists of nothing but knowledge workers, he knows that, in the long run, it will be far more valuable to him to have the people who developed the new product become part of Microsoft.

When Microsoft buys a competitor, it's not the product of the moment they are after, nor the information it takes to manufacture that product; *it's the people that produced the product.* What Gates is really buying is not a company, nor information, nor a product; it's more of the thing on which his huge business empire rests: knowledge. The easiest, and possibly the only, way to acquire that crucial knowledge is to acquire the people in which the knowledge resides.

On the other hand, the Microsoft approach is not all plain sailing. Acquiring knowledge through a company takeover is not without its problems. Difficulties—and even disaster—await to trap the unwary. Thinking of launching a takeover? Before you do, read on. . . .

## KNOWLEDGE ACQUISITION: THE FINE PRINT

If a company sets out to acquire the knowledge of another company by a corporate takeover, it had better be quite sure exactly where the key knowledge is located in the target company. Finding that out can be difficult, but getting it wrong can be very costly. In particular, if the acquisition involves some downsizing of the newly acquired company, as is often the case, then there is the very real danger of losing the very stuff that motivated the takeover in the first place: knowledge.

A good example of this is described by Davenport and Prusak [1998, p. 55]. In 1988, EL Products, a manufacturer of electroluminescent lamps, bought out a competing firm, Grimes. One thing that EL sought to gain by the takeover was Grimes's greater expertise in the efficient production of high-quality lamps. Unfortunately, what EL did not realize was that the key to Grimes's better production lay not in its managers or engineers, who were included in the takeover, but in the workers in its production line, who were let go. As a result, the acquisition did not give EL the knowledge it thought it was buying.

Even if a takeover does not include downsizing of the acquired organization, there is still the danger of a damaging clash between the management styles or the working cultures of the two companies. For one reason or another, the employees of the company being taken over might not like some aspect of the way the new parent company goes about its business. Some key employees may decide to leave, others may feel so disgruntled that they are no longer prepared to go that extra yard that was the key to the earlier success of their formerly independent company.

Of course, an obvious way to try to avoid the problem of clashing cultures is for the acquiring company to make a serious attempt not to tamper with the inner workings of the new acquisition, leaving all the existing management structure in place. But even then, things can go wrong. The employees of the acquired company may resent being part of the new parent company, and some key players may leave for that reason alone. No matter how gently the parent company proceeds, any takeover has an element of "conquerors" and "vanquished," and those in the acquired group may decide to take their talents elsewhere.

Still another possible pitfall of a takeover is that the parent company, used to its familiar ways of doing business, may resist being given advice from the new acquisition. This was a contributing factor to the failure of AT&T's purchase of National Cash Register some years ago. The telecommunications giant bought NCR in an attempt to break into the computer business. The acquisition turned out to be a complete failure. Part of the problem was undoubtedly the fact that NCR's general-purpose computer business was itself in bad shape prior to the acquisition. But it was also the case that AT&T was simply resistant to making the changes needed to become a computer company.

### SUMMARY

Information can be stored in books, manuals, databases, and in other ways. Knowledge, in contrast, only exists in a (human) mind—the mind of the *knower* of that knowledge.

Since information depends crucially on contexts and constraints, turning information into knowledge involves recognition of—and familiarity with—the relevant contexts and mastery of the appropriate constraints.

So-called stores of knowledge are really stores of information that has been chosen and structured by an expert so that a person with suitable training can readily internalize it and turn it into knowledge.

The easiest and often perhaps only way for one company to acquire the knowledge of a rival is to acquire the company—to buy the people in whose minds that knowledge resides, together with the working culture that makes that knowledge effective.

But any takeover is fraught with dangers:

- It can be hard to identify exactly where in the about-to-be-acquired company the key knowledge is located.
- There may be a clash of the management styles and working cultures of the two companies, leading to the departure of some key employees of the newly acquired company.
- The new parent company may resent "being told what to do" by the new acquisition, despite the fact that its original intention was to acquire that very knowledge that it now resists.
- Even if all of the above pitfalls are avoided, there remains the aftertaste of "conquerors and vanquished," and that alone can destroy the very thing the acquiring company set out to obtain: knowledge (which leaves with departing, disgruntled employees).

# 17

# In the Knowledge Game, People Beat Computers

## THE SEARCH FOR SILICON EXPERTS

During the 1980s, a considerable research effort was launched to try to develop computer systems that possessed (specialized) knowledge; in particular, computer systems that could provide "expert advice" in some specific domain. The motivation seemed reasonable enough: Even if computers cannot be programmed to exhibit the overall level of intelligence of a person—and despite what you may have read in books and magazines, no computer program has ever come remotely close to that goal, and probably never will—then perhaps it is possible to capture the very specific knowledge that a human expert brings to a particular task.

Such computer programs were called expert systems, and attempts were made to develop expert systems to provide medical diagnoses given data of symptoms, to locate oil or mineral deposits given topographic data, and to configure large computer systems to meet the needs of specific customers.

The idea was for an expert in the design of such systems—called, rather grandly, a "knowledge engineer"—to spend some time observing a human expert doing his or her job, and interviewing the expert to discover the different rules that the expert used to reach a conclusion. All of those rules were then coded in an appropriate programming language and fed into the computer, which was programmed to apply them in a systematic way to try to reach the desired diagnosis, prediction, or whatever.

The majority of such systems never came close to delivering the performance of a human expert, and the development of expert systems was all but abandoned. Almost none of the systems developed in the heyday of expert systems research is currently in use.

In Chapter 20, I shall provide a classification of expertise into five stages: novice, advanced beginner, competent, proficient, and expert. The classification is based on recognition of—and familiarity with—the relevant contexts and mastery of the appropriate constraints. In terms of the classification, the best performance achieved by an expert system was generally little more than stage 3, that of competence. Competence is the highest level of performance that can be achieved by following rules. Proficient and Expert performance, which are the areas where we see human *expertise*, cannot be completely captured by rules.

In terms of our present discussion of the distinction between knowledge and information, we would say that computer systems can at most process information, whereas true, human expertise requires knowledge. The human expert moves beyond competence by internalizing the stage 3 information and thereby converting it to knowledge.

## MOVING AT THE SPEED OF KNOWLEDGE

As the history of expert systems technology indicates, for all their uses in the Information Age, computers cannot replace humans when it comes to the application of knowledge. Given the phenomenal growth in both the speed and the capacity of computers over the past forty years, their inherent limitations are easily overlooked. The occasional apparent success of a computer system as a *knowledge* processor can sometimes lead to an altogether unachievable expectation that a knowledge problem can be solved by means of technology. This kind of success is so rare—and when it does happen it does so for very special reasons—that we need to be constantly on guard against making such an assumption.

For example, in the case of expert systems, achieving the level of "competence" was occasionally sufficient to meet the intended purpose. Thus, in some sense, there were some successful expert systems. For instance, the system to configure computer systems that I alluded to briefly above did all that was expected of it, and the company that developed it, Digital Equipment Corporation, used it successfully for several years. However, the reason for the success was that the task of configuring a computer system was sufficiently "mechanical" that rule following could be just as good as having a human expert perform the task. Moreover, even that system was eventually discarded, when the cost of continually updating the rules to accommodate the ever more rapidly changing components it was supposed to handle became prohibitively large.

Another example was the success of IBM's purpose-built chess computer, Deep Blue, which beat world chess champion Garry Kasparov in 1997. Deep Blue worked, not by developing expertise, but by brute force: It simply examined billions of sequences of future moves to see which one was the best. (Human chess players don't work like this at all.) Chess is sufficiently well-defined (each piece has a small number of precisely stipulated permitted moves) that such an approach can work. The challenge to Deep Blue's developers was to cope with the enormous numbers of possibilities that must be examined at each stage. To try to reduce that number, Deep Blue did have some rudimentary "expertise" built in—it was not *just* brute force—but again nothing beyond the "competence" level.

Knowledge can travel fast. Because people operate at the level of knowledge (as opposed to the more cumbersome data or information), we can change our behavior with great rapidity, often much more rapidly than a machine can be reconfigured or a computer reprogrammed.

For example, at its factory in Honjo, Japan, the electronic giant NEC has been steadily removing the robots from its assembly lines and replacing them with human workers. Why? Because the company has observed that the intelligence and flexibility of humans makes them far better equipped to cope with change than an automatic assembly line.

To take a specific instance, the company introduces a new model of mobile phone roughly every six months, so the cost of changing over to the production of a new model is a significant factor. The last time the company introduced a new model, humans reached target efficiency after making 8,000 units, but it took the production of eight times as many units (that is, 64,000 units) before the robot line met the target level. What is more, when both had reached peak efficiency, the people were 45 percent more productive than the robots. Instead of having to spend almost $10 million to change the production line for a new model, which was the case with the machines, a model change cost less than $2 million when the manufacture was carried out by people.

Armed with information, computers can be fast; operating with knowledge, people can sometimes be faster.

## KNOSMOSIS

When the expert systems researchers tried to endow computers with "expertise," they did so by trying to feed in the expertise one rule at a time. Provided there are enough rules, their argument went, the computer would eventually start to exhibit expert behavior. However, this is not the way people acquire expertise. Though rule acquisition can play a part in becoming an expert at something, for the most part the process is one of *adaptation*. By talking with, observing, or working alongside an existing expert, or by repeatedly trying to perform some task, a human can acquire knowledge and expertise by a process reminiscent of osmosis, the biological process whereby a cell absorbs nutrients through the cell wall. I like to refer to the acquisition of knowledge in this manner as "knosmosis."

An excellent illustration of knosmosis occurred in the course of one of the many attempts to develop an expert system. A major oil company wanted to develop an expert system that would scan aerial photographs of landscapes to locate the most promising sites to drill for oil.

The company hired a leading knowledge engineer to develop the system. He spent many months working with the company's top photo analyzer, a man who had proved himself extremely skilled at identifying potential drilling sites from aerial photographs.

By asking the photo analyzer to provide detailed explanations of the analytic process that led to each prediction, the knowledge engineer attempted to capture the analyzer's expertise in rules that could then be expressed in a form the computer could apply. Of course, the attempt was a failure. The expert system that was produced did not even come close to the degree of expertise of the human.

By the end of the project, however, the knowledge engineer had himself become highly skilled at analyzing aerial photographs to predict likely drilling sites. Indeed, it was generally acknowledged that, after the company's own photo analyzer, the knowledge engineer was the second-most-skilled person in the world at performing the task!

The attempt to capture the photo analyzer's expertise in a computer system was unsuccessful, but the process of interacting with the analyzer to try to codify his knowledge did result in the knowledge engineer achieving a high level of expertise. In effect, the process of observing the photo analyzer at close range provided the knowledge engineer with an apprenticeship in geological photo analysis, a modern-day version of the traditional system whereby craftsmen of times gone by passed on their knowledge and skill to the next generation.

It is important to realize that the knowledge engineer did not achieve his photo-analyzing expertise by learning all the rules that the human expert articulated. After all, the computer had all those rules, and yet it was not able to achieve acceptable results. Moreover, since the rules captured the information that the photo analyzer articulated and communicated to the knowledge engineer (and hence the computer), it would not be accurate to say that the knowledge engineer acquired his new skill by way of acquiring information. Rather, it was the very human process of observing and interacting with the photo analyzer that led to the knowledge transfer—*knosmosis.*

Of course, most successful managers know instinctively that the most effective way to acquire knowledge is by interaction with another person. Studies of managers in action have shown that they get around two-thirds of their knowledge from face-to-face meetings or telephone conversations and only one-third from documents or computers.

In the following chapter, we'll look further at the way people acquire knowledge and the role technology can play in facilitating knowledge transfer. For we are now ready to take a look at one of the hottest and fastest-growing areas of interest in management science: knowledge management.

---

### SUMMARY

In the 1970s and 1980s, expert systems were developed in an attempt to duplicate the knowledge and skills of a human expert, by capturing the human expert's knowledge and skills as a series of rules and programming a computer to reason with those rules.

Except in a small number of very special cases, expert systems did not achieve their goal. In terms of the knowledge versus information distinction, the explanation is that computers process information (strictly speaking, representations of information, but let's be generous for the moment), whereas genuine, human expertise involves knowledge.

Not only can knowledge lead to expert behavior, it can also enable its possessor to adapt rapidly to changing circumstances.

Humans often acquire knowledge by a process called "knosmosis." Knosmosis can occur when two people interact with each other over a period of time or when one person observes another performing a particular task many times.

Managers acquire roughly two-thirds of their knowledge through direct interaction with another person and only one-third from documents or computers.

# 18

# Where Can I Find Out About That?

--------------------------------------------------

## A NEW SPECIES IS BORN

With the growing realization that knowledge is the key not only to their future success but probably also their survival, many companies are now trying to improve the way they acquire, create, and transfer it. Knowledge management is the name of the game. Knowledge managers are showing up across the corporate world—although the term may become redundant, since smart companies have realized that in order for knowledge to reach all those parts of the organization that need it, *everyone* has to be a knowledge manager. Some of the larger companies, particularly in North America, have decided that the proper management of knowledge is so important that it needs to be put under the direction of a single, high-ranking person, usually called a "chief knowledge officer," or CKO for short (obviously designed to signify a rank comparable to that of the chief executive officer, or CEO).

In this chapter, we'll take a look at some of the challenges that face anyone charged with managing a company's knowledge flow. As we'll see, those challenges are so great that if a company seriously intends to make significant improvements in its knowledge flow, the person in charge definitely needs all the power and status of a CEO.

Individuals who are appointed to the position of CEO are typically drawn from the company's library and information services department, from the information technology department, or from the human resources department. The cases we shall examine will also make it clear that, in order to be successful, the CEO needs to break free of any one of those more established functions. For the task of knowledge management (that is, the task of organizing and directing the company's management of its knowledge) requires skills in all those areas and more.

## COMPUTERS WILL NOT SOLVE THE PROBLEM

Given the size and complexity of many of today's companies, effective knowledge management would not be possible without computers—and information technology in general. Thus, the chief knowledge officer needs to be familiar with the capabilities of the latest information technology.

But in the final analysis, computers have little to do with increasing knowledge or knowledge flow. In particular, in the absence of any other changes, the introduction into a company of a fancy new computer system or a networked suite of new desktop PCs will not lead to an improvement in the company's effective knowledge base.

Those "other changes" that have to be made are in the areas of management practices and company culture. Thus, the CKO needs to be able to manage and influence changes in those areas as well.

In fact, the introduction of information technology can hinder the flow of knowledge, as anyone who has spent any length of time working with computers will attest.

Even if the company escapes that all-too-familiar fate, there is still the very real danger that the sight of all the gleaming new hardware and colorful computer screens lulls the CEO into thinking that they have finally got the "knowledge problem" licked and hence need do nothing more than let the technology do the rest.

## . . . BUT THEY (COMPUTERS) CAN HELP

What information technology can contribute to successful knowledge management is support for interpersonal *communication*, particularly in a large company where face-to-face communication is often not possible. As a general rule, when a company depends on considerable tacit knowledge, technology should be used, not to try to encode the knowledge, but to help share that knowledge *directly*.

For example, British Petroleum (BP) uses computer videoconferencing technology to sustain what it calls "Virtual Teams." Introduced in 1994, Virtual Teams are groups of experts, at different locations, who maintain regular contact and work together to share their knowledge, helping one another with advice on how to solve problems.

What makes the Virtual Teams work is that the company views the fancy technology, not as an information system, but as a rich medium of communication. Human relationships are the key to the program's success, and BP goes to some length to ensure that the individual teams have every opportunity to bond into a collaborative group with a sense of common purpose. The regular videoconferencing meetings are backed up by periodic meetings in person, the company having realized that technology can never substitute for genuine face-to-face contact but can at best be used to sustain, for a period, relationships that have already been established.

In other words, the success of BP's Virtual Teams lies largely in the fact that the teams themselves are not at all virtual, they are real teams. The word "virtual" in the title refers to the use of technology to maintain regular contact between team members when they are separated physically. In setting up the program, BP's management realized that knowledge originates and resides in people's minds, and that the sharing of that knowledge for the company's good required mutual trust on the part of the team members. There had to be suitable (real) rewards for sharing knowledge, and real resources had to be provided to provide the conditions under which knowledge sharing could take place.

## THE OCCASION WHEN VIRTUAL PRESENCE SAVED THE DAY

On at least one occasion, the communications technology BP introduced to support its Virtual Teams had an unexpected payoff in terms of extending the physical reach of company knowledge.

In 1995, on an oil-drilling ship that BP was using to search for oil in the North Sea, the failure of a key, sophisticated piece of machinery brought all operations to a halt. Had it not been for the new information technology introduced to support the Virtual Teams, the company would have had to either fly a skilled engineer out from the mainland by helicopter or else send the ship back to port, both expensive options with a full crew sitting idle and a leasing cost of $150,000 per day for the ship.

However, on board the ship was a tiny video camera that could be linked via satellite to the company's offices in Aberdeen. This setup enabled a shore-based engineer to examine the faulty part visually, while talking to the onboard personnel by ship-to-shore phone. In this way, he was able to diagnose the fault and guide the crew through the necessary repairs. The entire episode lasted only a few hours, and drilling was then continued where it had been left off.

Throughout the repair process, the crucial knowledge remained on the shore, in the mind of the expert engineer. The technology allowed that knowledge to be applied many miles away on board a ship at sea.

## YELLOW PAGES FOR KNOWLEDGE

Some companies, realizing that most (if not all) of their knowledge lies in the minds of their employees, are developing what they call "knowledge maps" or "yellow pages." These are guides that lead the user, not to some body of information, but to the person or persons inside the company who have the particular knowledge sought.

Such guides to knowledge are generally implemented electronically, but the idea is not new: For generations, all journalists have had Rolodex cards containing names, addresses, and phone numbers for the many different experts they might need to consult in order to write a particular subject.

The card index works well for an individual user. But in order to provide all the employees in a large company with a yellow pages guide to the entire organization, technology is required.

For example, since 1995, Microsoft has been developing an electronic knowledge guide called the Skills Planning *"und"* Development system (SPUD). The system classifies each employee by his or her particular expertise, according to a classification scheme the company worked out to meet its needs. The idea is that a manager or an engineer who needs a particular collection of skills and knowledge can query the system for the names of the people who best meet those needs. The query can be restricted to Microsoft employees located in a particular part of the country or the world where the company has staff.

SPUD uses proprietary software, but many organizations use freely available World Wide Web technology to provide their members with a knowledge map. In those cases, a standard Web browser such as Netscape is used to navigate the guide. An example of such a system is Connex, a SPUD-like knowledge map that Hewlett-Packard developed for its R&D labs.

In many cases, the Web pages provided by knowledge maps include not only the name of the relevant expert and a brief account of the person's area(s) of expertise—the information that would obviously *have* to be provided by a knowledge "yellow pages"—but also a photograph, address, phone number, e-mail address, and sometimes a live video of the activity at the individual's workspace. These additional features can increase the effectiveness of such guides significantly.

## WHY THE PIF FEATURES ARE NOT JUST PIFFLE

One knowledge map that my colleague Duska Rosenberg and I have been examining is the People and Information Finder (PIF). The PIF is a Web-based, interactive knowledge resource developed by a consortium of European construction companies and universities for use in the construction industry. (The project is called CICC—Collaborative Integrated Communications for Construction.)

The modern, large-scale construction industry is an excellent example of an enterprise in which the use of multimedia communications technology can have a major impact. A present-day, large-scale construction project can take many months, or even several years, to complete, with different people being involved at different stages of the construction. As a result, at any one time, the different team members may be spread all over the world in different time zones. This can make it difficult to contact a relevant individual. For example, by the time the construction crew starts to pour concrete for a new shopping mall in Kent, England, the architect may have moved on to a project in Sydney, Australia.

PIF provides a common framework in which each member of the organization enters his or her own information. The use of a common layout means that each PIF Web page looks the same, so using the system quickly becomes routine. On the other hand, allowing each person to enter his own information gives him an opportunity to express his own personality. Significantly, each person is called the "owner" of the PIF page that displays his information.

Though it provides various kinds of information, the main purpose of the PIF is twofold: First, it provides a guide to lead users to the appropriate expert, whom they can then contact directly. Second, the PIF provides users with as much social background about that expert as possible, in order to increase the chance of a successful transfer of knowledge when that contact is made.

# David Leevers

Manager
BICC MultiMedia Communications
Group, Quantum House
Maylands Avenue, Hemel Hempstead
Hertfordshire HP2 4SJ
United Kingdom
Tel: 44 1442 210 100
Sec: 44 1442 210 125
Fax: 44 1442 210 101
Email: DavidLeevers@compuserve.com

24/12/97, 08:47

You might be able to catch a glimpse of me at work in the colour Video Glance above (click on "Reload/Refresh"), or in our Video Open Plan office, or see things from my point of view, literally.

## Nearest Neighbours

Take a look at our MMCG glance page, and you'll be able to see where all the members of the group are right now.

18/01/98, 00:55          19/01/98, 00:18          13/01/98, 12:12

Simon Soper * Nicholas Farrow * Bob Hamor

- Simon Soper - MultiMedia Communications contact for the MICC project
- Nicholas Farrow - CICC and VirtuOsi social worlds
- Bob Hamor - Technical and Support Contact
- Mark Wilderspin-Administrative contact for the RESOLV project
- David Sandys- Administrative contact for the CICC project
- David Hogg -Leeds University, Technical Representative for RESOLV project
- Tom Fernando - Ove Arup representative to the CICC project
- Mike Tubbs - Technical Manager - BICC
- Family members - an Altavista web search for Leevers

## Projects and Activities

1995 - Project Manager for European Commission (ACTS) Research projects

- RESOLV - REconstruction using Scanned Laser and Video
- CICC - Collaborative Integrated Communication in Construction

1993 - 1996 - Partner - EPSRC - DTI CSCW Project

- VirtuOsi - Support for Virtual Organisation, exploring the uses of Collaborative Virtual Environments in business and industry.   Partners include Lancaster Uni, Nottingham Uni, GPT, BT Labs, and Division

1992 - 1995   -   Partner- EC RACE project

- BRICC - BRoadband Communications in Construction. Encouraging the use of emerging communications services in the construction sector. The MultiMedia Hard Hat was first prototyped in BRICC. Development continues within the MICC project.

## Papers

- Inner Space - the Final Frontier - updated 22 August 1997
- 2.5D versus 3D - Guardian article by Keith Devlin, April 97
- Cycle of Cognition- A Fractal Framework for Social Cognition, Feb 96
- A Virtual Environment to support Multimedia Networking - July1993

A panoramic image of my office can be viewed online

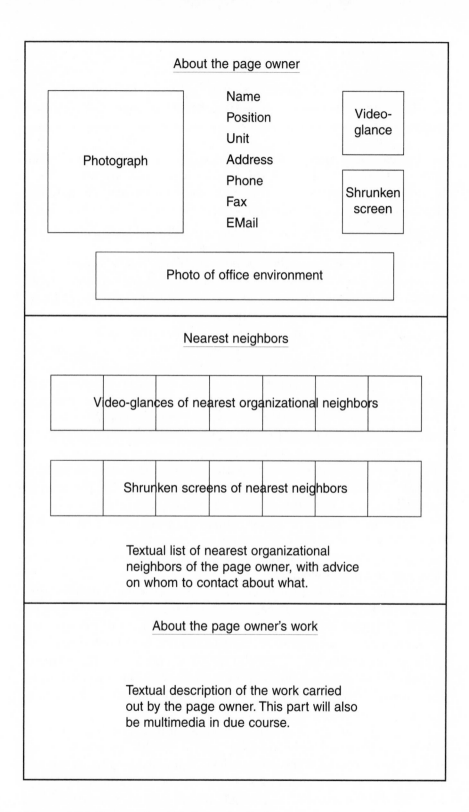

**Figure 18-1** The people and information finder.

A sample PIF Web page is shown in Figure 18-1 on pages 170 and 171, alongside a key to the layout. Each PIF page is divided into three main areas. The first (top) area, headed "About the page owner," provides information that helps a user locate the page owner in physical space: name, address, phone and fax numbers, and email address. (Mouse-clicking on the email address puts the user directly into an email window addressed to the page owner.) In addition, there is a portrait photograph of the owner, a live video of the owner's workplace, and a 360-degree photograph of the owner's office as seen from the owner's main desk chair.

The second region of the PIF page is called "Nearest Neighbors." It provides information about the page owner's organizational space, giving names of others in the organization who are engaged in similar or related work, together with Web links to those experts' own PIF pages. This enables users to access possible alternative sources of information if the page owner is not currently available.

The third and final area of the PIF page, called "Projects," describes the page owner's past and current projects, together with any other pertinent information the owner wishes to provide. Owners typically provide Web links to various other Web sources they think may be of benefit to a user.

The PIF is designed to make it clear that its main purpose is to put people in contact with one another. By far the greater part of the implementation effort for the PIF system, and by far the greater bandwidth required to bring a PIF page to a user's screen, are devoted to the conveyance of social information—of providing the PIF user with a sense of the page owner as a person. The goal is to prepare the way for a useful human-to-human contact between the PIF user and the page owner.

For instance, taken together, the video, the owner's photograph, and the 360-degree workplace photograph provide *human and social information* that enables a PIF user to form a mental image of the page owner as a person. Studies of electronically mediated communication (including the PIF itself) have indicated that this image is extremely valuable in helping to build the trust that is so important for knowledge transfer, particularly if the PIF inquirer and the page owner have never met.

In particular, the 360-degree photograph of the page owner's office, taken from the owner's chair, allows the PIF user to see what it is like to sit where the owner does—almost literally, to see the world through the owner's eyes. Is the owner surrounded by books and technical reports? How much technology is immediately to hand? Is he or she excessively neat or mildly scruffy? Initial studies of the PIF have shown that users rapidly form a mental image of the person they want to contact, before such contact is made, and that this initial image can condition their approach and thus increase the possibility of a successful knowledge transfer.

One important observation: In order to be effective, an organizationwide system such as SPUD or the PIF has to be made accessible to all. Such a knowledge directory is very different from the journalist's card index. Even if the present-day journalist implements her knowledge directory on a computer, she will very often try to keep its contents secret. It takes time for a journalist to build up an adequate network of contacts. Since journalism is a business that puts great emphasis on "being first with the story," there is considerable disincentive to giving away what becomes one's lifeblood. But most industries don't operate like that; they prosper by sharing their knowledge, as we shall see in the following chapter.

## SUMMARY

Many large companies are creating the position of a chief knowledge officer to oversee the management of knowledge.

Though a CKO needs to be familiar with current information technology, knowledge management is ultimately not about technology, but about people, management practices, and workplace culture.

The most significant contribution that technology can make to knowledge management is communicative: putting people in touch with one another.

Videoconferencing can maintain the cohesiveness of teams that are spread across different locations. Occasionally, it can extend an expert's reach, enabling him or her to work by telepresence.

Computer "knowledge maps" or "yellow pages" can help employees to identify and contact the person who has the special knowledge they need.

The inclusion in knowledge maps of portrait photographs, workstation videos, and photographs of the immediate workplace environment can help provide the inquirer with a mental image of the person she or he wants to contact. This social, background information can set the scene for an effective subsequent knowledge transfer.

# 19

# The Boston Beer Party and Other Tales

## IF WE ONLY KNEW WHAT WE KNEW

Lew Platt, CEO of Hewlett-Packard, once said, "If HP knew what HP knows, we would be three times as profitable." What Platt meant was that, in the minds of its many hundreds of talented employees was an immense amount of knowledge that was of potential use to HP, but much of that knowledge was not being used because it never found its way to the people who could take advantage of it.

Finding ways to ensure that the right knowledge finds its way to the person who can make good use of it is a major challenge facing any newly appointed chief knowledge officer.

The solution is not to be found in technology, as I argued in the previous chapter. Rather, the CKO's first step should be to develop a culture of knowledge creation and knowledge sharing. Since knowledge is a personal thing, something in the mind of each employee, knowledge creation and transfer can only occur if the cultural climate supports it.

One of the best examples of a firm that has successfully developed such a culture is 3M. That company sees itself as a product innovator. With 60,000 different products on the market, some

30 percent of the company's revenues come from products less than four years old. In 1996, the company brought out 400 new products. Its goal is to have at least 10 percent of its profits in any year come from products less than a year old.

Innovation on that scale could not happen without a strong culture of knowledge creation and knowledge transfer. Every 3M employee is encouraged to come up with ideas for new products, and everyone can apply for company grants to support the development of a new idea.

It was in this culture that sandpaper salesman Dick Drew was able to invent Scotch tape and have the company bring it to market. In many organizations, he would have been unable to get anyone to listen to his idea. After all, isn't product development the province of the Product Development Department? Surely, salespeople are not supposed to waste their time thinking about new products; their task is to sell the company's existing line! In many companies such sentiments may prevail. But not at 3M.

## THE BOSTON BEER PARTY

Even when management is serious about knowledge transfer, achieving the desired results can be difficult. Moving large amounts of information from one part of the organization to another is relatively easy. Transferring even a small amount of knowledge can prove much more problematic.

A case in point arose during the construction of the Boston Harbor Tunnel a few years ago.

The main contractor had carried out a similar project in New Zealand, constructing a tunnel linking two islands. The tunnelers who had worked on the New Zealand project had developed some innovative drilling methods that the company wanted the Boston workers to adopt.

Company managers tried various means to transfer details of their new methods from New Zealand to Boston: descriptions, diagrams, instruction manuals, and the like. Nothing worked. The Boston workers seemed unable—or unwilling—to adopt the new ideas. In the end, the frustrated company decided that the only thing left to try was to fly some of the New Zealand tunnelers to Massachusetts to meet the Boston workers face to face. When they did that, evenings spent together in a local bar drinking beer achieved what no amount of formal communications could do. By talking man to man with the New Zealand tunnelers, the Boston workers came to understand the new methods—how to use them and why they worked—and were able to employ them on the American project.

There were two reasons why the company resisted this solution for so long. First, there was the cost of bringing the New Zealand workers halfway around the world and putting them up in Boston for several days. Companies often fly senior executives around the world, but not manual workers.

The second, and far more significant reason was cultural. The managers of the construction company were all engineers, used to dealing with codified information in the form of instruction manuals and blueprints. They felt that it had to be possible to achieve the desired knowledge transfer in that manner; indeed, that was the only means they knew. What they failed to appreciate was that it was not an issue of *information* transfer but *knowledge* transfer. Information transfer can lead to knowledge transfer in the right culture, provided the two parties already have sufficient relevant knowledge. But in the absence of those necessary prerequisites, face-to-face conversation may be the only solution.

Although the company's engineers were used to acquiring knowledge from information in documents, the Boston tunnelers worked in a culture where skills were passed on from person to person or acquired by experience. No amount of documentation from a group of unknown "foreigners" from the other side of the world was going to influence the way they did things. On the other hand, when they got to know the New Zealand workers, they saw that they were "regular guys," very much like themselves, having similar skills and experiences. At that stage they were open to suggestions, and were prepared to trust the advice they were given—advice given in the form of first-person "war stories" to which they could easily relate on the basis of their own experiences.

## GETTING ALL STEAMED UP AT MOBIL

In the case of the Boston Harbor Tunnel project, knowledge transfer via documentation (that is, via information) failed in large part because the tunnelers did not work in a culture where knowledge came from manuals. But even where experts do work with "blueprints and manuals," an established working culture can often generate a considerable resistance to knowledge transfer.

A case in point is what happened when engineers at a Mobil Oil site in Kansas developed a new way to determine how much steam was required to drill under different conditions. They tried out the new method and the results were dramatic, giving rise to major cost savings.

Eager to achieve similar efficiency gains at all their sites, the company asked the Kansas engineers to write up the details of the new method, together with an account of the success rate they had achieved, and sent it to all their other drilling sites. What was the result? Not one of the other sites so much as tried the new method.

Puzzled, the company decided to adopt more persuasive means of getting the message across. Consultants were bought in to carry out a case study of the Kansas field and write a report that could be circulated to the other sites. Videos were made of the designers of the new process, and those too were circulated to the other sites. Time was set aside at each site for a discussion of the pros and cons of the new approach.

After six months of such effort, some 30 percent of the other sites had adopted the new process. So some progress had been made. On the other hand, for 70 percent of the sites, the well-established fact that the new method would result in an immediate, significant increase in efficiency still did not appear sufficient to persuade the site engineers to change their approach.

What was the cause of such seemingly irrational resistance to change? Almost certainly, the following two factors played a major role.

First, the Mobil engineers at the various sites had, over the years, learned to trust their own methods—methods that had served them well. As engineers, they were used to acquiring knowledge from documents, of course. But the process of turning documented information into knowledge was carried out within the framework of their existing knowledge. Being urged to modify that framework, even by means of videos of other engineers just like themselves, was not sufficient to persuade them to change. The only way to do that might have been to bring some of the Kansas engineers to the other sites, in much the same way that the New Zealand tunnelers had been flown to meet the Boston workers.

But in the Mobil case, even face-to-face meetings might not have worked. Engineers tend to be fiercely independent operators, who are used to relying on their own skills and experience—generally with good reason. That independence was particularly strong at Mobil. Moreover, the corporate culture within Mobil included a distrust of bragging, and that too provided a major force acting against adoption of the new method.

The experience at Mobil shows that it can be difficult, if not impossible, to transfer knowledge if the company culture acts against such transfer. Unfortunately, changing a well-established culture can take a long time. It often requires the retirement of a significant portion of the existing personnel. It is much easier to build such a culture from the outset, as was the case with 3M, described earlier.

## KNOWLEDGE IS POWER

No matter how much commitment a company's management has for the free flow of knowledge, in the end it comes down to individual human behavior. If knowledge is an important asset to the company—and for the majority of companies it is—then the possession of crucial knowledge confers power on its possessor. That power may take different forms, real or imagined. In some instances, knowledge can lead to the power to influence events on a daily basis. Alternatively, the possession of key knowledge might simply provide an individual with potential leverage against being dismissed if the company has to downsize. A third possibility is for the possession of knowledge to provide someone, not with real power, but merely with a sense of self-importance. Whichever possibility is the case, a person in a company who has crucial knowledge will generally feel some incentive to maintain exclusivity on that knowledge. Practically anyone who works in even a moderately large organization will have come across the person who is always reluctant to give up the information he or she possesses.

Countering this natural human tendency is no easy task. Financial or other incentives can help, though they need to be more than just tokens. But a major factor is the organizational culture. For an example of the way that a well-established culture of sharing led to a highly successful knowledge transfer, let's go back to 3M.

The invention of Post-it Notes by 3M's Art Fry came about as a result of a memo from another 3M scientist who described the new glue he had developed. The new glue had the unusual property of providing firm but very temporary adhesion. As a traditional bonding agent, it was a failure. But Fry was able to see a novel use for it, and within a short time, Post-it Notes could be seen adorning practically every refrigerator door in the land.

Yet Post-its could never have happened if 3M had not developed a strong culture of sharing ideas. Unlike many companies, at 3M is easy for any employee to acquire information from others in the company. For instance, the firm organizes regular meetings and "knowledge fairs" where its employees are encouraged to share their knowledge. In addition, aware that not all seemingly good ideas will work out in practice, 3M has demonstrated an understanding tolerance for mistakes. Obviously, this kind of culture cannot simply grow from the bottom up. It has to begin with a clear message from the top levels of management, who must then support and encourage the growth of that culture of sharing at all organizational levels.

And yet, so problematic is the issue of knowledge transfer that, even when a firm has the kind of enlightened management of 3M, there are at least two major obstacles that can prevent the company from benefiting from much of the knowledge of its workers.

The first obstacle is the question of who benefits from that knowledge. Many highly ambitious engineers guard their growing knowledge and expertise until they think the time is right to make real use of it, and then they leave the company to form their own startup company. This is, of course, the history of most of the computer companies in Silicon Valley. (The history of the others is that their founders dropped out of graduate school at Stanford in order to capitalize on the knowledge they had acquired while there.) In such circumstances, there is little the company (or the university) can do to make use of that knowledge itself, even if it knows where it is located.

The second obstacle is best exemplified by what happened at Xerox PARC. Xerox established its Palo Alto Research Center (PARC) in order to carry out leading-edge research into new communications technologies. The enterprise was one of the greatest success stories of modern times, with the highly talented researchers at PARC developing most of the features of modern desktop computing, including the graphical user interface with its windows and mouse, and the Ethernet that connects computers together.

For all the advances made in the lab, however, repeated attempts by the PARC directors to persuade Xerox to take advantage of the new ideas met with complete indifference. The Xerox management was familiar with the world of xerography—so dominant was the company that it had invented the very terminology of the industry—but knew little about computers and did not appreciate the enormous explosion in personal computer use that was about to take place. As a result, they simply were not interested. It was left to the young Stephen Jobs, co-founder of Apple computer, to see the potential in the exciting new approach to computing that PARC had developed and make use of it in the Macintosh computer, which took the world by storm soon after its introduction in 1984.

## THE BEST KNOWLEDGE TRANSFER ENVIRONMENT ON EARTH

The story of Xerox and Apple is a classic case of knowledge transfer from one company to another. It is also one of the most dramatic, with a well-established and powerful company allowing the transfer to a brash young upstart company of one of the greatest product innovations of the twentieth century. But the Apple Macintosh is just one of literally hundreds of new technology products that have come out of California's Santa Clara Valley—or Silicon Valley, as it is now more widely known.

Why has Silicon Valley been so successful at innovation, and why have attempts to duplicate that success elsewhere met with far less success? For example, consider Route 128 in Boston; Austin, Texas (where Sematech and MCC are based); Research Triangle Park in North Carolina; and Silicon Glen in Scotland. Each of these locations has a similar infrastructure to Silicon Valley: Each has one or more major universities in the vicinity, and most offer a lifestyle that can attract young, smart individuals to come and live. (If the San Francisco peninsula seems more attractive as a place to live than any of the others, the cost of housing surely mitigates against that advantage.)

Among the factors put forward to explain the success of Silicon Valley, one that is surely highly significant is the well-established culture of sharing ideas. Walk into any coffee shop or restaurant in Palo Alto—the acknowledged capital of Silicon Valley—and you will be surrounded by talk of leading-edge technology.

Of course, the successful, large computer companies try to maintain a veil of secrecy around their development projects, a goal that can in large part be achieved by offering their employees first-class dining and recreational facilities on site. But the researchers and developers who work for the smaller, younger companies—companies that they very often co-own—exchange ideas freely. Among that section of Silicon Valley there is a very strong culture of knowledge sharing.

Silicon Valley's strong tradition of knowledge sharing is greatly assisted by the tremendous mobility in the industry. Many of the leading-edge researchers in Silicon Valley have worked for several of the major companies and often a whole chain of small startups. Their current company allegiance may be only a few months old and at most a couple of years. Their expectation is that in a short while they will be working somewhere else on a different project. When asked where they work, they are more likely to answer "Silicon Valley" than to give the name of their current company. And it is not too far from the truth to say that that is where they see their professional identity, and where their allegiance lies—in the society and culture that is Silicon Valley.

Sure, the year-round California sunshine, the beautiful surroundings, and the proximity of San Francisco all help to draw people to Silicon Valley. But the secret to the Valley's success in the rapidly changing world of computer technology is the culture of sharing that pervades the region.

Though attempts to duplicate Silicon Valley elsewhere have largely failed, the underlying framework of knowledge sharing can be developed within a single organization anywhere in the world. But a manager or CKO who wishes to develop a strong culture of sharing within his or her own company will not succeed unless he or she has a deep appreciation of the crucial importance of human relationships and the development of trust.

## HOW CAN YOU RECOGNIZE A GOOD IDEA?

There is one further aspect of Silicon Valley that almost certainly contributes to its success. That is the almost total lack of any status-based intellectual hierarchy. In Silicon Valley, you are as good as your latest idea. This almost anarchic culture leads to the extremely flat management structures for which Silicon Valley companies are famous.

There is a natural tendency in all of us to assign value to information based on the status of its source. For example, CEOs often look no further than their own vice presidents for advice, despite the fact that, particularly when technical issues are involved, far better advice is almost certainly to be found much lower down the company hierarchy.

Similarly, pronouncements from academics at Harvard or Princeton tend to be taken more seriously than statements emanating from Penn State or the University of Texas. And yet the United States has enjoyed an oversupply of highly qualified Ph.D.'s for many years, with the result that you can find academics at practically any major university who are just as able as those at the Ivy League schools.

As is clear from the description I gave earlier, 3M managed to avoid the "status trap." But they are very much the exception. Can other companies follow 3M's lead? One obvious way to encourage innovation is to eliminate evaluation based on an assessment of the innovator. How? Set up a system whereby new ideas are evaluated by a team that is not informed of the source—so-called blind evaluation. But this method doesn't always work. In the university world, when faced with considerable evidence that referees were evaluating research papers based largely or solely on the address of the author, some research journals tried to adopt a process of "blind refereeing," but this met with strong resistance. One obvious reason for the opposition was that it made refereeing much more difficult. Another equally obvious reason was that it meant that far more papers from authors at prestigious institutions would be rejected! The example of the universities does not mean that blind evaluation of ideas cannot be made to work. But it does indicate that it might require considerable effort to meet with any success.

### SUMMARY

Knowledge creation and knowledge transfer can take place only in a supportive culture and in a manner with which the participants are familiar.

For some groups, face-to-face contact may be the only means to achieve effective knowledge transfer.

Employees generally develop their own ways of doing things, and it can be difficult to persuade them to modify their approach. A strong, logically sound argument might not be enough.

It is also important to remember that knowledge confers power on its owner. Thus, there is an incentive for individuals to hoard the knowledge they have. This is particularly true for individuals who feel that their positions are threatened.

The best way to support knowledge transfer is to develop a strong institutional culture of knowledge sharing, supported by a good knowledge-sharing infrastructure of meetings, knowledge fairs, retreats, development grants, and so on, and a tolerance of mistakes. It has to be accepted and understood that anyone can have a good idea, regardless of status.

The case of Xerox PARC highlights the importance for senior management to be open to new ideas and changing markets.

Try to avoid the human tendency to evaluate information in terms of the status of the person who delivers it. Information should be measured in terms of its content.

# 20

# Becoming an Expert

- - - - - - - - - - - - - - - - - - - - - - - - - - - - - - - - - - - - - - - - - - - -

## HOW DO YOU TRAIN AN EXPERT?

How do you master new techniques or become adept at using new skills? How do you—or how does your company—train employees to become expert in performing their duties?

The situation-theoretic view of information developed in Chapters 1 through 6 of this book has implications not only for communication but also for expertise—for what it means to be an expert and how expertise can be acquired. Those implications apply as much to sports and other recreational pursuits as they do to expertise in the workplace.

In fact, situation theory's implications for expertise are essentially the same as the results of an analysis of expertise carried out some years ago by the philosopher Hubert Dreyfus and his brother Stuart, an engineer, and presented in their book *Mind Over Machine*. In some respects, the similarity is hardly an accident. The Dreyfuses' work was one of the factors I took into account in developing a situation-theoretic account of expertise in the first place. On the other hand, it says a lot for both the fundamental ideas of situation theory and the work of the Dreyfuses that an analysis of expertise based on situation theory leads naturally to the Dreyfuses' account.

I'll begin with a description of the five stages of expertise, as proposed by the Dreyfuses. From there I'll indicate how this taxonomy follows naturally from a situation-theoretic view. Incidentally, the

Dreyfuses based their analysis on a study of the skill-acquisition process of airplane pilots, chess players, car drivers, and adult learners of a second language.

The discussion of expertise will provide us with a powerful illustration of the fact that I have mentioned several times already—and which has been demonstrated by practically every real-life example we have looked at—namely, that turning information into knowledge is largely a matter of recognizing and becoming familiar with the relevant contexts and mastering the appropriate constraints. In particular, it is *not* about "learning the information" in the narrow sense of memorizing the data. In many cases, that is certainly an important first step in the acquisition of knowledge (or skill). But it is just that: a first step.

## THE FIVE STAGES OF SKILL ACQUISITION

*Stage 1* of skill acquisition is what the Dreyfuses called the novice stage. A novice approaches the activity by following rules, which he or she follows in an unquestioning, context-free fashion. For example, in the case of learning to drive a car with a manual transmission, a novice driver follows rules such as "change gear at such-and-such a speed" or "follow the car in front at such-and-such a distance." Novice performance is generally easily recognized as such. The novice car driver's movements are sudden and jerky, the changing-gear rule is applied rigidly, with no account taken of contextual factors such as the sound of the engine or the degree of slope and other road conditions, and the driver tries to maintain the recommended separation distance regardless of the traffic density.

*Stage 2* is called advanced beginner. The feature that distinguishes the advanced beginner from the novice is that, while both act in a rule-following fashion, the advanced beginner modifies some of the rules according to context. For example, the advanced beginner car driver will take account of engine sound in deciding when to change gear, and will adjust the separation distance from the car in front according to traffic density. In situation-theoretic terms, the advanced beginner has started to recognize certain types and modify the rules according to those types.

*Stage 3* is called competence. The competent performer still follows rules but does so in a fairly fluid fashion—at least when things proceed normally. Instead of stepping from one rule to another, making a conscious decision of the next step at each stage—behavior characteristic of the novice and advanced beginner stages—the competent performer has a much more holistic understanding of all the rules. He has an overall sense of the activity and chooses freely among the rules for the appropriate one. For example, the competent driver drives with a particular goal in mind and pays attention to engine sound and traffic density in making choices of gear and vehicle position. However, since the act of driving still occupies his full attention, he gives little thought to passenger comfort, road courtesy, or even safety and the law. Moreover, he is unlikely to be able to respond well to a sudden emergency.

*Stage 4* is proficiency. For much of the time the proficient performer does not select and follow rules. Rather she has had sufficient experience to be able to recognize situations as being very similar to ones already encountered many times before, and to react accordingly, by what has become, in effect, a trained reflex. For instance, the proficient driver may realize, quite subconsciously, that she is approaching a sharp corner too fast, given the rainy conditions, and so may decide to ease off on the accelerator or apply the brakes. Though the action of slowing down or braking involves making a conscious decision and following a rule, the driver comes to make that decision as a result of a quite unconscious instinct, based on past experience of similar circumstances.

*Stage 5* is that of the expert. The expert performer does not follow rules and indeed is generally not consciously aware of any rules governing the activity. Rather she performs smoothly, effortlessly, and subconsciously. The expert driver is not aware of the car she is driving, and she may not even be consciously aware of her driving. When things proceed normally, the expert performer does not make decisions, follow rules, or solve problems; rather she simply *does what normally works.*

## THE TRUE EXPERT

According to the above account, a genuine *expert* is not someone who has simply learned to follow the rules in a highly efficient and rapid fashion. Rather, an expert *is not following rules at all*, not even subconsciously. Rules are there to help us learn how to perform a task, but once we become expert at performing that task, we no longer need the rules. According to this view, rules are like the training wheels we sometimes use in order to help young children learn to ride a bicycle. At first, the training wheels are set to be in contact with the ground, where they provide constant support. A little later, when the child has had some practice, the training wheels are set a little higher, so the child can learn to feel what it is like to ride without the constant support of the wheels but can still be protected from the danger of falling over. Then, finally, when the child has learned how to ride the bicycle, we remove the training wheels altogether. At that stage, the child's skill has eliminated any need for the training wheels.

In many ways, the spectrum of levels of expertise is parallel to (and is related to) our data-information-knowledge classification. One classification applies to activities, the other to the possession of information. Stage 1 of expertise corresponds roughly to information so simply and directly linked to its representation that it could almost be classified as data. Stages 2 and 3 of expertise correspond more or less to the possession of information. Stages 4 and 5 correspond to knowledge.

In situation-theoretic terms, expertise amounts to type-driven activity. The expert acquires the ability to recognize certain types and to respond with an action of an appropriate type. That is, the expert develops certain constraints linking types of situations encountered and types of actions performed. Proficiency—Stage 4 in the above hierarchy—represents the stage in which the individual has formed the type-recognition ability and the requisite constraints but does not automatically apply those constraints.

With its emphasis on type recognition and constraints, situation theory is ideally suited to handle expertise. In contrast, other rule-based approaches are limited to the first three stages of expertise. This explains why the field of expert systems never delivered on its original promises. As a branch of artificial intelligence, expert systems research tried to use formal logic to encode expertise into computer systems. But true expertise is not rule-based. As the Dreyfuses captured well in their taxonomy, genuine expertise arises when the rules are dispensed with, and such behavior cannot be encoded into a computer program (at least, not a traditional, rule-based computer program).

## HOW TO DEVELOP EXPERTISE

As any manager knows, true expertise only comes through *experience,* an observation reflected in the common root of the words "expert" and "experience." Reading books and attending training seminars are valuable, but they can only take you so far. Would you want to entrust your medical care to a "physician" who had learned everything from books and lectures but had no experience in clinical practice? There is a good reason why the medical profession insists that all new doctors undergo an extensive internship during which they can practice real medicine on real patients under close supervision.

Likewise, the training of airline pilots comprises both theoretical instruction in the form of lectures and the acquisition of knowledge from books, together with many hours of practice in a flight simulator, capped off with a period of closely supervised real flying.

Similarly, many other professions require a period of on-the-job training prior to certification.

The inadequacy, on its own, of rule-based instruction is something we all become aware of when we first learn to ride a bicycle or learn to swim. Despite being told how to do it, the only way to learn is to keep trying—to go through that awful period of repeatedly falling off the bicycle or sinking beneath the water—until that moment when, as if by magic, we suddenly "get it." And as we all know, once we have acquired such a skill, we almost never lose it.

The same is true for taking on a new job. The skillful manager knows that a new employee, or an existing employee beginning a new job, needs time to acquire the necessary skills. During that time, the new person will almost certainly make mistakes. If the job is a critical one, where mistakes can be extremely costly, then ways have to be found to provide the necessary experience through some kind of simulation: closely supervised internship for a new physician or a period in a flight simulator for a new airline pilot.

The training of managers can present a particular problem, since, in many businesses, one of the most important skills is to be able to cope with the new and the unexpected. How can you provide people with experience at handling situations that have never occurred before—nor perhaps even been envisaged? This is where true expertise—Stage 5 in the Dreyfuses' taxonomy—really differs from the earlier stages of skill acquisition. New situations are almost always a novel combination of familiar circumstances. It is a fortunate feature of human beings that in a domain where they have expertise, they are often very quick to adapt that expertise to deal with novel situations.

For example, experienced airline pilots do encounter new situations from time to time, sometimes highly threatening ones, and they generally cope with them extremely well. Similarly, an experienced physician, when faced with an unfamiliar combination of symptoms, will often come up with the correct diagnosis.

The greater the variety of circumstances that are provided in the initial on-the-job training period, the more likely the individual will be to be able to cope with new situations. Indeed, true Stage 5 expertise can be characterized as the level of expertise in which a person is able to cope with novel situations within the domain of expertise.

This is why businesses are increasingly turning to various kinds of simulation to train their managers. One method involves setting up teams of managers that compete against one another in trying to solve some artificial problem. Another approach uses actors to play the roles of customers or competitors in staged simulations of business encounters. Such training methods are as yet hardly a "science," and many people in the business world are highly skeptical of this kind of approach. But the lesson to be learned from the study of information and expertise on which this book is based is clear: Apart from the often highly costly method of jumping in the deep end and learning from actual mistakes, various kinds of simulation are the only possible way to develop genuine management expertise. The scientific reason is the nature of types, an issue we take up in the next chapter.

## SUMMARY

We introduced a scale showing five different levels of skill: novice, advanced beginner, competent, proficient, and expert. The first three involve explicit—and usually conscious—rule-based behavior; the fourth involves fairly fluid behavior that makes use of rules, but in an automatic and generally unconscious manner; the fifth level does not use rules, rather behavior is instinctive.

The possession of skill can be compared with the possession of information and knowledge: The first skill level corresponds to the possession of data, skill levels 2 and 3 correspond to the possession of information, and skill levels 4 and 5 correspond to having knowledge.

For many activities, such as performing mathematics and driving a car, initial learning can be achieved by means of rules, by being told or shown what to do in a step-by-step fashion. On the other hand, sometimes rules are of minimal help, and the only way to learn is by practice. Swimming or riding a bicycle are examples of activities learned in such a way. But, however the early learning is achieved, increased expertise, up to level 5 on the scale, can then come about in just one way: as a result of practice and experience.

The fact that true expertise can only be achieved through practice has major implications for the way managers and other professionals are trained. You have to find a way to provide a person with sufficient—and sufficiently broad—experience. That experience can be provided either on the job (ideally under supervision by an expert) or by means of simulations. To be effective, simulations have to be as realistic as possible.

# 21

# Why Expertise Cannot Be Taught

## Rules Aren't Enough

In the previous chapter, we observed that, to recast it in situation-theoretic terms, expert behavior does not mean simply efficient rule following but rather type-driven activity. That is, the expert acquires the ability to recognize certain types (often types of situations) and to respond with actions of appropriate types—the expert develops certain constraints linking types of situation encountered to types of actions performed.

We first met the concept of a type in Chapter 4, where we saw how types play a crucial role in the storage and transmission of information. The example of wearing slippers indicated how our behavior is influenced by types. The approach to expertise adopted in this book extends that treatment of patterned behavior by treating expertise as automatic and fluid responses to certain types of situations.

So what is the difference between the expertise of a surgeon with twenty years of experience and a person who knows when it is appropriate to wear slippers? In terms of the cognitive mechanisms involved, the answer is: no difference at all. The distinction lies in the number of types (of objects, situations, actions, and so on) involved and the degree to which those types are differentiated.

The person who knows when it is appropriate to wear slippers and when it is not does not follow rules. He or she simply recognizes two types of situations: the type when it is appropriate to wear slippers and the type when it is not. Each of those two types is associated—in an automatic, unconscious way—with a type of behavior: wearing slippers and not wearing slippers. Which type of situation is applicable to which type of behavior? Well, it's virtually impossible to give a precise answer. Certainly, "within one's own home" is generally thought of as one type of situation for which it is appropriate to wear slippers. But this is not always the case. For example, wearing slippers is not appropriate when the boss is coming for dinner, or if the house is under 6 inches of floodwater, or if the carpet fitters are laying a new carpet and there are sharp nails sticking up everywhere, and so on. And it's that "and so on" that prevents you from being completely precise.

Likewise, "being out of the home," including being at work, is generally thought of as a type of situation in which it is not appropriate to wear slippers. But there are plenty of exceptions. For example, it is permissible to wear slippers out of the home if you recently had a sports injury in which you damaged your foot, or if . . . well, why don't you fill in the next two or three examples? And when you have provided those additional examples, don't forget the "and so on."

The point of the above discussion, of course, is to indicate that, even for very simple, everyday examples, it is usually impossible to say precisely what are the types that govern our behavior. The only way to pin down the type of situation in which it is appropriate to wear slippers is to refer to it that way, that is, as the type of situation in which it is appropriate to wear slippers! Since practically every one of us is an expert in knowing when it is appropriate to wear slippers, that way of describing the type of situation is perfectly adequate. But it does not tell you exactly what the type consists of!

So how does a surgeon describe—precisely—the conditions under which to perform a certain operation? If it were possible to provide a complete and accurate description, then it would, in theory, be possible to learn to perform an operation by simply reading an instruction manual. But few of us would entrust our operation to a person who had just read the books. We would want someone with *experience,* and the more the better.

## THE EXPERT TYPE

Expertise consists primarily of the ability to recognize types. In the case of a surgeon, for example, the number of types (of ailments, symptoms, and so on) that the expert needs to be able to recognize is fairly large, and the distinction between them can be quite fine. (This is one reason for specialization, for having heart surgeons, brain surgeons, kidney specialists, and so on.) With each type recognized, the expert needs to associate it with an appropriate type of behavioral response, but that association should come automatically once the type of situation has been recognized. The real key to expertise is the ability to recognize the relevant types (of ailments, symptoms, and so on).

Once the expert has acquired the ability to recognize a sufficient number of types, it then becomes possible to differentiate them into new categories—in effect, create new types. One way this can be done is by refining an existing type, for example, refining the categorization of a type of flu to the type of a particular strain of flu. Another way to create a new type is by splitting an existing type into two or more subtypes. For example, the type of cholesterol splits into the type of high-density lipids and the type of low-density lipids. Or we can combine types to create new types. A physician will combine the type of a smoker and the type of a person who is overweight to give the type of a person who is both a smoker and overweight. Such a person might receive different advice than that received by a person who is just overweight.

One of the key features of a true expert is the ability to create new types that no one has recognized before; for example, the physician who notices a novel combination of symptoms and thereby discovers a new kind of illness.

In many cases, it is possible to give fairly comprehensive descriptions of a type, and this is the key to instruction. For instance, it is possible to enumerate conditions under which it is normally safe to land a jumbo jet. Not only can these conditions be taught to beginning pilots, they can be programmed into a computer to create automatic landing systems. Problems arise when conditions are not quite the same as those in the list or when other circumstances interfere.

For example, what happens if an earthquake produces a large crack in the runway but leaves all the automatic guidance systems in operation? An automatic landing system might well continue to try to land the plane, but even an inexperienced pilot would know at once that it is no longer safe to land. All of the other conditions that normally constitute the type when it is safe to land would prevail, but the large crack in the runway results in a different type, when it is not at all safe to land.

The reason expert systems technology cannot produce genuine expertise, and the reason it would be foolhardy in the extreme to dispense with pilots on airplanes, no matter how reliable we think the control systems are, is that there is always the possibility of the unforeseen circumstance (such as an earthquake). By and large, humans have evolved to respond quite well to novel circumstances—that's one reason why we are here and some other species is not—but machines can't do that.

To give you some idea of why expertise—in particular, efficient type recognition of the appropriate kind—has to be developed through experience, and cannot be acquired just by following rules, let's take a look at a familiar and, on the face if it, very simple type: the type of a seat.

## WHAT IS A SEAT?

Forget all the secondary meanings of the word. Just think of "seat" as the everyday variety, the kind of thing you sit on. How would you describe the *type* of a seat?

How about this: Four legs, a horizontal flatish piece you can sit on, and a back? That's a good start, but what about three-legged seats? Or six-legged? Or single-legged? Come to that, it might have no legs at all—the base could be an enclosed cupboard. How flat is the bit you sit on? How big? What shape? What exactly constitutes a back? What about a stool that has no back?

After some thought, you realize that seats can come in all sorts of shapes and sizes, with no legs or any number of legs from one to maybe seven or eight, possibly more. What about a car seat? A seat in a train? An airplane seat? Surely a simple garden bench consisting of a plank of wood resting on two large slabs of stone is a seat.

Let's look at the question of size. Does the type of a seat include small chairs designed for young children? What about a studio prop of an enormous chair used in one of those movies where a person is shrunk to the size of an insect? It's way too large for anyone to sit in it, but is it still a seat?

Okay, what if we define a seat differently? Instead of trying to specify its shape or size or construction, let's look at its function. Let's describe a seat as something we sit on. But we sometimes sit on the ground. Does that mean the ground is a seat? Surely not. Let's exclude the ground. On a picnic, we sit on a blanket. Does that make a blanket a seat? How about a miniature seat for a doll's house? Is that a seat or not? Or consider a very valuable antique chair that has become so fragile that no one can sit on it. Do we want to exclude that from being classified as a seat? How about a film-prop chair made of balsa wood, designed to be broken over a stuntman's head in a fight scene? It is not designed to be sat in, but does that make it not a seat?

By now it should be fairly obvious that we could go on like this indefinitely. No matter how you try to pin down exactly what constitutes a seat and what does not, it is always possible to think of exceptions—things that satisfy your specification but "obviously" are not seats and things that do not satisfy your specification but "obviously" are seats. And yet none of us feels the least bit uncertain using the word "seat." We all feel we know what a seat is. In short, the *type* of a seat is a definite concept in our cognitive world.

How did we ever acquire the concept of a seat? Clearly not by being given a definition—we have just seen how no definition could capture our concept of a seat. Rather, we each developed the concept—the *type*—through our day-to-day experience in the world. We each became an expert in recognizing the type "seat." The fact that we are unable to provide a complete and accurate specification of what a seat is does not for one moment prevent us from accurately and easily classifying any object we come across as a seat or a nonseat.

The same is true of many of the other types we meet and use in everyday life. For instance, try specifying what is meant by the word "game." How about the type "bird"?

Once you have convinced yourself that there is no alternative to experience to properly acquire some of the "simple" types we meet and use in our everyday lives, ask yourself what provisions are provided for your employees, colleagues, or managers to acquire the types they need simply to carry out their jobs reasonably well, let alone to become stage 5 experts at the job. The fact is, it is not enough just to provide them with information, no matter how much you throw at them. Some of that information might be invaluable. But for expert performance, no amount of information can take away the necessity for experience.

## SUMMARY

It can be difficult and often impossible to provide a complete and accurate specification of even very common types such as "seat." Yet human beings can acquire and use types as part of their conceptual framework. Such acquisition comes through experience, over time.

Being aware of the nature of types and the way people learn to recognize and use them is an important asset for any manager. Expertise can only be achieved by partaking in enough experiences to acquire and recognize the requisite types.

# 22

# The 5-Percent Solution

---

## THE GOLDEN RULE

In the early chapters, we developed a way of thinking about information. By introducing the concepts of situations, types, infons, and constraints, we were able to produce a theoretical framework to analyze the way information is represented, acquired, processed, and transmitted. In subsequent chapters, we used that framework to look in detail at some specific examples taken from the real world of business.

What is the practical goal? By understanding information—that is, *information* as opposed to representations, or big-I information rather than little-i information—we hope to find ways to improve information flow. In the Knowledge Society, it is imperative that we learn how to make sure that the right information gets to the right people at the right time in the right form.

There are, of course, many important factors that affect the flow, accuracy, and usability of information, and the more factors we take into account, the better our information systems will be. But if I had to distill from our investigations a single slogan that, if followed, would have the greatest positive impact on information management—personal or in business—I would have no difficulty. It's this:

*Context matters.*

The situation-theoretic view of information takes this slogan into account from the get-go, in the very notion of how information is assumed to arise: namely, an object or configuration of objects in the world represents information by way of certain constraints in a suitable context. Without the context, *there is no information.* In different contexts, the same object or configuration of objects can represent quite different information. This is exactly what occurred with tragic results for Flight AA 965 (discussed in Chapter 7), when the letter R represented two different directional beacons in two different contexts.

Other contextual influences on information flow that we looked at are human psychology (particularly commitment) and cultural/social factors.

The powerful effect of psychological context is utilized by the skillful salesperson who begins a transaction by showing us an item he or she knows we do not want and almost certainly won't buy, but compared to which the item the salesperson really wants to push will seem cheaper and/or better. Indeed, if the salesperson is really lucky (or extremely good at the sales job), we might even be persuaded that we have been offered an incredible bargain that is too good to miss.

Commitment is an equally powerful force in setting a context for future action, as the experiment with the "environmental" billboards in California showed.

Somewhat similar to commitment is our natural inclination to repay debts to others. Have you ever wondered why charities keep sending you supplies of return labels with your name and address printed on them, or why Hare Krishna collectors at airports begin by offering you a "free" flower? It's because (as has been verified by numerous experiments) most of us find it hard not to respond by making a donation, even if—as is usually the case—we don't want the gift and know when we accept it that we will throw it away immediately. Once again, our actions are strongly conditioned by the context.

Besides exerting a strong influence on our actions, social and cultural factors play a particularly significant role when it comes to turning information into knowledge. Knowledge is information possessed in a form that makes it available for immediate use to guide action. Turning information into knowledge is not just a matter of "memorizing the facts." Rather, it also requires learning to recognize and become familiar with the relevant constraints and achieving mastery of the appropriate constraints. Indeed, "memorizing the facts" is generally the easiest part of the process of acquiring knowledge. (This is why we say that knowledge can only be found "inside people's heads," and not in books, manuals, reports, and charts.)

Time and again, the key to success in industry has not been skill, education, or information. In fact, it's probably accurate to say that most competing firms of roughly the same size have fairly comparable employees. Rather, in many cases, the crucial X factor is the company culture: Does it encourage, support, value, and reward innovation and knowledge sharing? (Not just in its slogans and advertising literature, but in its day-to-day practice.)

One thing surely is clear: Whichever you look at—information, knowledge, psychology, commitment, social factors, or culture—context matters.

Does the analysis of contexts developed in the early chapters provide us with any tools to help us deal with context? This is the question we turn to next.

## IT CAN BE WORTH A THOUSAND WORDS

In Chapter 7, we introduced a simple way of representing contexts diagramatically, to provide a geometric view of logical reasoning. In Chapters 8, 10, and 11, we used context diagrams to analyze the structure of a conversation (calling them conversation diagrams in that case).

Of course, drawing a diagram is not in itself a magic cure-all. For one thing, it doesn't help you to decide what the important contexts are. Moreover, you still have to understand the domain sufficiently well to figure out what the crucial constraints are. Using a conversation diagram, however, does two things for us: First, it reminds us of our golden rule: Context matters. That alone is sufficient reason to take such diagrams seriously. Second, a conversation diagram provides a systematic way to *represent* the key contexts *once you have identified them*.

The second of the above two benefits is a real one that we shouldn't underestimate. The underlying rationale behind the use of such diagrams is to take advantage of the tremendous power of visual representations as aids to comprehension and to reasoning. Line diagrams are particularly effective in that regard. As the nineteenth-century logician John Venn showed with his Venn diagrams, simple geometric representations—based on a natural, spatial metaphor—can help to clarify our understanding of a complex situation. (Venn diagrams represent complex logical situations by means of inter-secting ovals, much like our use of context diagrams to represent conversations.)

The reason for the effectiveness of diagrams is not hard to uncover. In human evolutionary history, the recognition of spatial configurations came long before the ability to use language or carry out logical reasoning, and for most of us a simple diagram is far eas-ier to comprehend than a piece of text or a sequence of mathemati-cal equations. Indeed, the understanding provided by a well-drawn diagram can often be instantaneous. That is why we so often say that "A picture is worth a thousand words." (Figure 7-3 on page 78 is surely a dramatically clear way to indicate the problem that caused American Airlines Flight 965 to crash.)

## THINK LIKE A GENERAL

Although we introduced the context diagram as an analytic device, there is nothing to prevent us from using one to plan a real conversation, much as a general might plan a military assault.

Planning a conversation with a context diagram forces us to view the exchange as a joint exploration of territory—specifically, an exploration that takes information in the individual cognitive terri-tories of the two participants and puts it into the shared common ground. In part, such an approach may be useful because it makes explicit a metaphor that is already present in language: namely, we speak of "steering the conversation," we say that a conversation "veered into unfamiliar territory" or "stuck to familiar territory," that the participants "stayed on familiar ground," or even, on occa-sion, that they "established common ground."

By way of a simple illustration, let's take another look at the case (described in Chapter 10) of the real estate agent who persuaded the elderly couple to sell their New England farm. Recall that the cou-ple had lived on the farm all their lives and consistently refused all offers to buy, claiming that they would never dream of leaving their lifetime home. A skillful real estate agent was able to negotiate the sale of the farm to his client by correctly reasoning that security was the couple's principal motivating factor. He was then able to lead them to make the decision that their security would increase if they sold the farm and purchased a new home.

Imagine that you are the real estate agent and you are trying to decide how to approach the elderly farmer. You might draw a

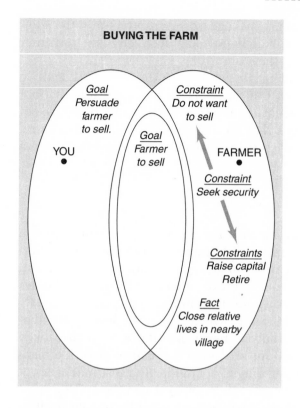

**BUYING THE FARM**

**Figure 22-1** Buying the farm.

conversation diagram as in Figure 22-1. The goal is to establish in the common ground an agreement for the farmer to sell his farm. So you begin by entering this goal in the common ground.

Since it is your goal to persuade the farmer to sell, you write down this constraint in your background situation. You also know that the farmer has consistently refused all offers to purchase the farm. You enter this constraint in the farmer's background situation.

As things stand, the farmer's "will not sell" constraint and your "I want him to sell" constraint are in total opposition. So you need to enter his cognitive territory via an alternative route that does not bring you up against his "will not sell" constraint. You reason that the "will not sell" constraint is probably a consequence of the "I want to be secure" constraint. (Notice that you are now reasoning about the information structure in the farmer's cognitive territory.) For elderly people especially, security involves two essential aspects: home and finances. From your discussions with the farmer's neighbors, you have also gleaned the information that the farmer's wife has a close relative in a nearby village, so you can factor in the additional security that comes from living near close relatives. You enter all of these constraints on the diagram.

Now you have your alternate route in to the farmer's cognitive territory: You concentrate, not on the issue of selling a home, but on having the financial security of money in the bank, being free of the daily financial worries that go with trying to maintain a farm, and getting the personal security that comes from living near a close relative.

## FISHBONE SOUP

One obvious drawback with using a context diagram to plan a real conversation is that, prior to the conversation, there are generally quite a number of contexts that may turn out to be important. With more than three or four contexts, a context diagram becomes impossibly complex. (After a conversation is completed, it is usually clear what the key contexts were, which is why a conversation diagram is more useful to analyze a conversation *after* it has taken place.) A far better method for analyzing contexts as part of a planning process is the fishbone diagram, familiar to most managers.

The fishbone diagram came into popular use in the 1970s as part of the "Deming management method," and rapidly proved itself to be a useful planning tool.* By changing the way the diagram is applied, we can use it to identify the different contextual factors that affect information flow.

Referring to Figure 22-2, you start the diagram by filling in the box on the right. This is the final informational state, S. This can be a state that already exists, for which you want to trace its predecessors; or it can be a state you desire, for which you want to plan a sequence of events leading to it.

Then you fill in the fishbone to the left. Each square represents a particular context that influences the path leading to the final state S. Against the lines leading from a context C to the main axis you note the names of the constraints in context C that play a role in achieving S. A plus sign (+) indicates a supporting constraint; a minus sign (–) an inhibitory constraint. The far left box (or boxes) are used to represent the social and cultural context.

In drawing a fishbone diagram to support the analysis of infor-

*See Walton (1986).

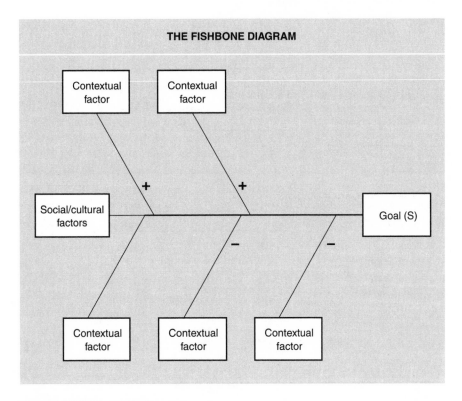

**Figure 22-2** The fishbone diagram.

mation flow, the aim is to identify the contexts that play a role and which you should take into account. Since human communication always takes place in a social/cultural context—a context that all the evidence shows is almost always significant—you should always include at least one social/cultural box. Thereafter, you should include as many context boxes as you can think of.

It may be that one or more of the lines leading from a box—say, E—to the main axis has no annotations denoting constraints. This does not mean that there are no relevant constraints—if there were no constraints, then the situation E would *not* be a context for achieving S. Rather, absence of a constraint denotation simply means that the analysis has not (yet) led to the identification of any constraint in that context.

Clearly, the fishbone diagram does not provide the useful territorial picture of the conversation diagram, but it does make very clear what are the critical contexts.

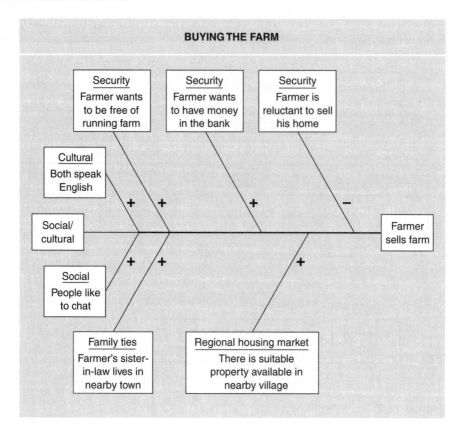

**Figure 22-3** Buying the farm.

Figure 22-3 gives a fishbone diagram for our example of the farmer and the real estate agent. Notice that a fishbone diagram does not tell you *how* each context will affect any subsequent course of events (if at all). Rather, the intention is simply to make you aware of the various contexts that *can* play a role. To be forewarned is to be forearmed.

## BUT IS IT SCIENCE?

For all that we can make practical use of some of our analytic tools, however, let me reiterate that the main goal of this book is to help us *understand* information and knowledge. Through understanding comes practical progress. (It is often said, with considerable justification, that there is nothing as practical as a good theory.)

One obvious benefit of our study is that it has uncovered (some of) the complexities inherent in the creation, representation, storage, and transmission of information. In addition, our study has highlighted the significant *human* aspect of knowledge.

In connection with that last remark, surely you will have noticed that the later chapters of the book are much more anecdotal than the earlier ones. Yes, the same underlying framework is present throughout: contexts (situations) and constraints. Moreover, since knowledge depends upon information, everything in the later chapters builds upon the material developed earlier. (You can't understand knowledge until you understand information.) But our approach is different when we focus on knowledge, because knowledge is much more complex than information. Information is amenable to a study that, if not entirely mathematical, has at least some of the flavor and rigor of mathematics. Not so for knowledge.

You may have noticed that I kept using the word "study" above. However, I claimed at the start of the book that it would be more than a "study." I said the aim is to present a *scientific analysis* of information. Has it fulfilled that promise?

In any analysis, you have to choose what to emphasize and what to ignore (or at least downplay). To make such choices, you need appropriate criteria. To ensure even a reasonable chance of success, these criteria should have a justifiable rationale. But once you have that, you have the beginnings of a science: You get a science (or the start of one) when you use a justifiable rationale to make choices as to which features will form the basis of your study and subsequent descriptions, and which will be ignored. In that respect, I would claim that the preceding pages do present, at the very least, the beginnings of a science of information.

For example, our theoretical analysis of conversations has at least some of the features of any mathematical science. We started out by viewing a conversation as a joint exploration of territory, as captured in the conversation diagram. The key contexts—the background situations, the common ground, and the focal situation—are regions of what we might call information space.

In a similar vein, the background situations might also be usefully called cognitive spaces. Each participant in the conversation operates according to the constraints within his or her own background and manipulates information within that background.

The purpose of a meeting can be regarded as the movement of items in the different background situations into the common ground. Such movement is caused by the participants jointly visiting that item in information space. A participant may take information within her own background and, by making a successful contribution to the conversation, put it into the common ground. (Making a statement that the others accept is an example of such a contribution.) Or she may use her words (and possibly other actions) to get another participant to take information from his background and put it into the common ground. (Successfully asking a question is an example.)

Is it (the beginnings of a) science? I would say it is. You may disagree. In any event, a more important question is: Is it useful? The applications of situation theory listed in the Prologue notwithstanding, I'll leave that question for you to answer. But to some extent, the answer probably depends on what you mean by "useful." Let me elaborate.

## THE 5-PERCENT SOLUTION

In the early chapters of this book, we looked closely at the structure of information. Using situation theory, we were able to bring to the study of information a degree of scientific precision that has been sorely lacking in most previous studies.

In the later chapters, starting with a firm understanding of information, we used examples to illustrate how to turn information into knowledge.

It is clear that the precision we achieved in the early chapters, while far greater than previous studies, is not on the same level as the mathematical precision characteristic of the natural and physical sciences. Such precision is almost certainly not possible when individual human behavior is the focus. The reason is that the methods of mathematics and science, so effective in studying the physical world, do not apply to individual human behavior—at least not to anything like the same extent.

On the other hand, such use of the methods of mathematics and science in the human domain is not in itself new. Economists, psychologists, sociologists, management scientists, and others have been blending the use of mathematical and scientific techniques with the specific methods of their disciplines for many years. (In *Goodbye, Descartes,* I referred to such uses of mathematics as "soft mathematics.")

Consider, for example, the use of the bell curve in all four of the disciplines with which the professionals mentioned above are involved: economics, psychology, sociology, and management science. The theory of the bell curve is classical, rigorous mathematics. When used in any one of those disciplines, it can lead to results that, while lacking the absolute precision of the natural sciences, are generally reliable and useful.

That last point is important, so let me take it a step further. The kind of scientific analysis presented in this book does not lead to results having the same "absolute certainty" that you can get when you apply the scientific method to the physical world. What it can do is give you results that are more reliable or more precise than you could obtain otherwise. But this may be all you need. In business, it is often enough to find "the 5-percent solution"—a 5-percent increase in output or efficiency that can be the difference between success and failure. The "science" described in this book may be able to give you that 5-percent solution.

## INTO THE FUTURE

Most (though not all!) of the results obtained so far by applying situation theory to the study of human activity have been either post mortem analyses conducted after something has gone wrong or else confirmations of known or at least suspected results. But as our understanding of information improves and more people make use of the developing new science of information, we can expect to see more results that are clearly new.

It is worth remembering that the development of a new science does sometimes follow in the heels of technology or other human innovation. Indeed, contrary to the popular conception, this is the way science normally proceeds! For example, steam engines were developed and used long before physicists produced a scientific theory that explained why they worked, and electricity was being created and put to use well in advance of the development of the corresponding science.

On the other hand, despite this similarity between the development of situation theory and the development of other new scientific theories, it should be borne in mind that applications of situation theory are not going to develop into another "precise" science. In most cases, the 5-percent solution is almost certainly the best that can be expected, or even hoped for. If what you want are complete solutions, you are destined for disappointment: Situation theory will not give you what you want, nor will anything else. There is no magic bullet. But, if you are in a business in which that 5-percent solution is significant—and that surely includes any enterprise that needs to make a profit in an open market—then you may find that the ideas presented in this book can be of real use.

## Summary

The conversation diagram provides a useful tool to visualize a two-person conversation in a "geographic fashion." In principle, it could be used as a planning tool, but its real benefit is that it provides an intuitive way to visualize a conversation, based on a powerful spatial metaphor.

For real-life planning, the fishbone diagram can be used as a practical tool to identify all the contexts that influence a particular instance of information flow.

When it comes to analyzing information flow, situation theory will not provide a complete solution. But it can give you the 5-percent solution, a 5-percent increase in efficiency or productivity. And in a competitive market, that is probably all you need.

# BIBLIOGRAPHY ----------------------

Aczel, P., Israel. D., Katagiri, Y., & Peters, S. (Eds.). (1993). *Situation theory and its applications* (Vol. 3). Stanford, CA: CSLI Publications, Stanford University.

Barwise, J. (1989). *The situation in logic.* Stanford, CA: CSLI Publications, Stanford University.

Barwise, J., & Etchemendy, J. (1987). *The liar: An essay in truth and circularity.* Oxford, UK: Oxford University Press.

Barwise, J., & Etchemendy, J. (1994). *Hyperproof.* Stanford, CA: CSLI Publications, Stanford University.

Barwise, J., & Perry, J. (1983). *Situations and attitudes.* Cambridge, MA: MIT Press.

Carroll, J. (Ed.). *Designing interaction: Psychology at the human-computer interface.* New York: Cambridge University Press.

Clark, H. (1992). *Arenas of language use.* Stanford, CA: CSLI Publications, Stanford University.

Davenport, T., & Prusak, L. (1998). *Working knowledge.* Cambridge, MA: Harvard University Press.

Devlin, K. (1991). *Logic and information.* Cambridge, UK: Cambridge University Press.

Devlin, K. (1996). *A situation-theoretic model of processes.* Preliminary report from the Advanced Research Projects Agency, Agile Manufacturing Pilot Program SOL BAA94-31A.

Devlin, K. (1997a). *Goodbye, Descartes: The end of logic and the search for a new cosmology of the mind.* New York: John Wiley.

Devlin, K. (1997b). *Using situation theory to compare the effectiveness of different office configurations.* Report prepared for Steelcase, Inc.

Devlin, K., & Rosenberg, D. (1993). Situation theory and cooperative action. In P. Aczel, D. Israel, Y. Katagiri, & S. Peters (Eds.), *Situation theory and its applications* (Vol. 3; pp. 213-264). Stanford, CA: CSLI Publications, Stanford University.

Devlin, K., & Rosenberg, D. (1996). *Language at work: Analyzing communication breakdown in the workplace to inform systems design.* Stanford, CA: CSLI Publications, Stanford University.

Dreyfus, H. L., & Dreyfus, S. E. (1986). *Mind over machine: The power of human intuition and expertise in the era of the computer.* New York: Macmillan, The Free Press.

Freedman, J., & Fraser, S. (1966). *Journal of Personality and Social Psychology.*

Gawron, P., & Peters, S. (1990). *Anaphora and quantification in situation semantics.* Stanford, CA: CSLI Publications, Stanford University.

Gumpertz, J., & Hymes, D. (Eds.). (1972). *Directions in socialinguistics: The ethnography of communication.* New York: Holt, Rinehart & Winston.

Karat, J., & Bennett, L. (1991). Working within the design process: Supporting effective and efficient design. In Carrol, J. (Ed.), *Designing interaction: Psychology at the human-computer interface.* New York: Cambridge University Press.

Menzel, C., & Mayer, R. (1996). Situations and processes. *Concurrent Engineering: Research and Applications,* 4(3):229-246.

Panko, R., & Kinney, S. (1992, January 7-10). *Dyadic organization communication: Is the dyad different?* Proceedings of the 25th Hawaii International Conference on Systems Sciences, pp. 244-253.

Penzias, A. (1989). *Ideas and information.* New York: Norton.

Rosenbery, D., & Hutchison, C. (Eds.). (1994). *Design Issues for CSCW.* New York: Springer-Verlag.

Sacks, H. (1972). On the analyzability of stories by children. In J. Gumpertz and D. Hymes (Eds.), *Directions in sociolinguistics: The ethnography of communication* (pp. 325-345). New York: Holt, Rinehart & Winston.

Schwartz, D. (1995). The emergence of abstract representation in dyad problem solving. *The Journal of the Learning Sciences, 4* (3): 321-345.

Senge, P. (1990). *The fifth discipline: The art and practice of the learning organization.* New York: Currency Doubleday.

Stasser, G. (1992). Pooling of unshared information during group discussion. In S. Worchel, W. Wood, and J. Simpson (Eds.), *Group process and productivity.* Newbury Park, CA: Sage Publications.

Stasser, G., Taylor, L.A., & Hanna, C. (1989). Information sampling in structured and unstructured discussions of three- and six-person groups. *Journal of Personality and Social Psychology,* 57, 67-68.

Taylor, L. A. (1990). *Program planning effectiveness: Improving the quantity and quality of idea generation with memory aids and structured interaction.* Ph.D. dissertation, Miami University, Oxford, OH.

Walton, M. (1986). *The Deming management method.* New York: Putnam.

Worchel, S., Wood, W., & Simpson, J. (Eds.).(1992). *Group process and productivity*. Newbury Park, CA: Sage Publications.

# INDEX